T0281696

Scalable Big Data Architecture

A Practitioner's Guide to Choosing
Relevant Big Data Architecture

Bahaaldine Azarmi

Apress®

Scalable Big Data Architecture

ISBN-13 (pbk): 978-1-4842-1327-8

ISBN-13 (electronic): 978-1-4842-1326-1

Managing Director: Welmoed Spahr
Lead Editor: Celestin Suresh John
Development Editor: Douglas Pundick
Technical Reviewers: Sundar Rajan Raman and Manoj Patil
Editorial Board: Steve Anglin, Pramila Balen, Louise Corrigan, Jim DeWolf, Jonathan Gennick,
        Robert Hutchinson, Celestin Suresh John, Michelle Lowman, James Markham, Susan McDermott,
        Matthew Moodie, Jeffrey Pepper, Douglas Pundick, Ben Renow-Clarke, Gwenan Spearing
Coordinating Editor: Jill Balzano
Copy Editors: Rebecca Rider, Laura Lawrie, and Kim Wimpsett
Compositor: SPi Global
Indexer: SPi Global
Artist: SPi Global
Cover Designer: Anna Ishchenko

Distributed to the book trade worldwide by Springer Science+Business Media New York, 233 Spring Street, 6th Floor, New York, NY 10013. Phone 1-800-SPRINGER, fax (201) 348-4505, e-mail orders-ny@springer-sbm.com, or visit www.springeronline.com. Apress Media, LLC is a California LLC and the sole member (owner) is Springer Science + Business Media Finance Inc (SSBM Finance Inc). SSBM Finance Inc is a Delaware corporation.

For information on translations, please e-mail rights@apress.com, or visit www.apress.com.

Apress and friends of ED books may be purchased in bulk for academic, corporate, or promotional use. eBook versions and licenses are also available for most titles. For more information, reference our Special Bulk Sales–eBook Licensing web page at www.apress.com/bulk-sales.

Any source code or other supplementary material referenced by the author in this text is available to readers at www.apress.com. For detailed information about how to locate your book's source code, go to www.apress.com/source-code/.

For Aurelia and June.

Contents at a Glance

Contents

About the Author

Bahaaldine Azarmi, Baha for short, is a Solutions Architect at Elastic. Prior to this position, Baha co-founded reachfive, a marketing data-platform focused on user behavior and social analytics. Baha has also worked for different software vendors such as Talend and Oracle, where he has held positions such as Solutions Architect and Architect. Baha is based in Paris and has a master's degree in computer science from Polyech'Paris. You can find him at linkedin.com/in/bahaaldine.

About the Technical Reviewers

Sundar Rajan Raman is a Big Data architect currently working for Bank of America. He has a bachelor's of technology degree from the National Institute of Technology, Silchar, India. He is a seasoned Java and J2EE programmer with expertise in Hadoop, Spark, MongoDB, and Big Data analytics. He has worked at companies such as AT&T, Singtel, and Deutsche Bank. Sundar is also a platform specialist with vast experience in SonicMQ, WebSphere MQ, and TIBCO with respective certifications. His current focus is on Big Data architecture. More information about Raman is available at https://in.linkedin.com/pub/sundar-rajan-raman/7/905/488.

Sundar would like to thank his wife, Hema, and daughter, Shriya, for their patience during the review process.

Manoj R Patil is a principal architect (Big Data) at TatvaSoft, an IT services and consulting organization. He is a seasoned business intelligence (BI) and Big Data geek and has a total IT experience of 17 years with exposure to all the leading platforms like Java EE, .NET, LAMP, and more. In addition to authoring a book on Pentaho and Big Data, he believes in knowledge sharing and keeps himself busy providing corporate training and teaching ETL, Hadoop, and Scala passionately. He can be reached at @manojrpatil on Twitter and writes on www.manojrpatil.com

CHAPTER 1

■ ■ ■

The Big (Data) Problem

Data management is getting more complex than it has ever been before. Big Data is everywhere, on everyone's mind, and in many different forms: advertising, social graphs, news feeds, recommendations, marketing, healthcare, security, government, and so on.

In the last three years, thousands of technologies having to do with Big Data acquisition, management, and analytics have emerged; this has given IT teams the hard task of choosing, without having a comprehensive methodology to handle the choice most of the time.

When making such a choice for your own situation, ask yourself the following questions: When should I think about employing Big Data for my IT system? Am I ready to employ it? What should I start with? Should I really go for it despite feeling that Big Data is just a marketing trend?

All these questions are running around in the minds of most Chief Information Officers (CIOs) and Chief Technology Officers (CTOs), and they globally cover the reasons and the ways you are putting your business at stake when you decide to deploy a distributed Big Data architecture.

This chapter aims to help you identity *Big Data symptoms*—in other words when it becomes apparent that you need to consider adding Big Data to your architecture—but it also guides you through the variety of Big Data technologies to differentiate among them so that you can understand what they are specialized for. Finally, at the end of the chapter, we build the foundation of a typical distributed Big Data architecture based on real life examples.

Identifying Big Data Symptoms

You may choose to start a Big Data project based on different needs: because of the volume of data you handle, because of the variety of data structures your system has, because of scalability issues you are experiencing, or because you want to reduce the cost of data processing. In this section, you'll see what symptoms can make a team realize they need to start a Big Data project.

Size Matters

The two main areas that get people to start thinking about Big Data are when they start having issues related to data size and volume; although most of the time these issues present true and legitimate reasons to think about Big Data, today, they are not the only reasons to go this route.

There are others symptoms that you should also consider—*type of data*, for example. How will you manage to increase various types of data when traditional data stores, such as SQL databases, expect you to do the structuring, like creating tables?

This is not feasible without adding a flexible, schemaless technology that handles new data structures as they come. When I talk about types of data, you should imagine unstructured data, graph data, images, videos, voices, and so on.

Yes, it's good to store unstructured data, but it's better if you can get something out of it. Another symptom comes out of this premise: *Big Data is also about extracting added value information* from a high-volume variety of data. When, a couple of years ago, there were more read transactions than write transactions, common caches or databases were enough when paired with weekly ETL (extract, transform, load) processing jobs. Today that's not the trend any more. Now, you need an architecture that is capable of handling data as it comes through long processing to near real-time processing jobs. The architecture should be distributed and not rely on the rigid high-performance and expensive mainframe; instead, it should be based on a more available, performance driven, and cheaper technology to give it more flexibility.

Now, how do you leverage all this added value data and how are you able to search for it naturally? To answer this question, think again about the traditional data store in which you create indexes on different columns to speed up the search query. Well, what if you want to index all hundred columns because you want to be able to execute complex queries that involve a nondeterministic number of key columns? You don't want to do this with a basic SQL database; instead, you would rather consider using a NoSQL store for this specific need.

So simply walking down the path of data acquisition, data structuring, data processing, and data visualization in the context of the actual data management trends makes it easy to conclude that *size is no longer the main concern*.

Typical Business Use Cases

In addition to technical and architecture considerations, you may be facing use cases that are typical Big Data use cases. Some of them are tied to a specific industry; others are not specialized and can be applied to various industries.

These considerations are generally based on analyzing application's logs, such as web access logs, application server logs, and database logs, but they can also be based on other types of data sources such as social network data.

When you are facing such use cases, you might want to consider a distributed Big Data architecture if you want to be able to scale out as your business grows.

Consumer Behavioral Analytics

Knowing your customer, or what we usually call the "360-degree customer view" might be the most popular Big Data use case. This customer view is usually used on e-commerce websites and starts with an unstructured clickstream—in other words, it is made up of the active and passive website navigation actions that a visitor performs. By counting and analyzing the clicks and impressions on ads or products, you can adapt the visitor's user experience depending on their behavior, while keeping in mind that the goal is to gain insight in order to optimize the funnel conversion.

Sentiment Analysis

Companies care about how their image and reputation is perceived across social networks; they want to minimize all negative events that might affect their notoriety and leverage positive events. By crawling a large amount of social data in a near-real-time way, they can extract the feelings and sentiments of social communities regarding their brand, and they can identify influential users and contact them in order to change or empower a trend depending on the outcome of their interaction with such users.

CRM Onboarding

You can combine consumer behavioral analytics with sentiment analysis based on data surrounding the visitor's social activities. Companies want to combine these online data sources with the existing offline data, which is called *CRM (customer relationship management) onboarding*, in order to get better and more accurate customer segmentation. Thus, companies can leverage this segmentation and build a better targeting system to send profile-customized offers through marketing actions.

Prediction

Learning from data has become the main Big Data trend for the past two years. Prediction-enabled Big Data can be very efficient in multiple industries, such as in the telecommunication industry, where prediction router log analysis is democratized. Every time an issue is likely to occur on a device, the company can predict it and order part to avoid downtime or lost profits.

When combined with the previous use cases, you can use predictive architecture to optimize the product catalog selection and pricing depending on the user's global behavior.

Understanding the Big Data Project's Ecosystem

Once you understand that you actually have a Big Data project to implement, the hardest thing is choosing the technologies to use in your architecture. It is not just about picking the most famous Hadoop-related technologies, it's also about understanding how to classify them in order to build a consistent distributed architecture.

To get an idea of the number of projects in the Big Data galaxy, browse to https://github.com/zenkay/bigdata-ecosystem#projects-1 to see more than 100 classified projects.

Here, you see that you might consider choosing a Hadoop distribution, a distributed file system, a SQL-like processing language, a machine learning language, a scheduler, message-oriented middleware, a NoSQL datastore, data visualization, and so on.

Since this book's purpose is to describe a scalable way to build a distributed architecture, I don't dive into all categories of projects; instead, I highlight the ones you are likely to use in a typical Big Data project. You can eventually adapt this architecture and integrate projects depending on your needs. You'll see concrete examples of using such projects in the dedicated parts.

To make the Hadoop technology presented more relevant, we will work on a distributed architecture that meets the previously described typical use cases, namely these:

- Consumer behavioral analytics

- Sentiment analysis

- CRM onboarding and prediction

Hadoop Distribution

In a Big Data project that involves Hadoop-related ecosystem technologies, you have two choices:

- Download the project you need separately and try to create or assemble the technologies in a coherent, resilient, and consistent architecture.

- Use one of the most popular Hadoop distributions, which assemble or create the technologies for you.

Although the first option is completely feasible, you might want to choose the second one, because a packaged Hadoop distribution ensures capability between all installed components, ease of installation, configuration-based deployment, monitoring, and support.

Hortonworks and Cloudera are the main actors in this field. There are a couple of differences between the two vendors, but for starting a Big Data package, they are equivalent, as long as you don't pay attention to the proprietary add-ons.

My goal here is not to present all the components within each distribution but to focus on what each vendor adds to the standard ecosystem. I describe most of the other components in the following pages depending on what we need for our architecture in each situation.

Cloudera CDH

Cloudera adds a set of in-house components to the Hadoop-based components; these components are designed to give you better cluster management and search experiences.

The following is a list of some of these components:

- **Impala:** A real-time, parallelized, SQL-based engine that searches for data in HDFS (Hadoop Distributed File System) and Base. Impala is considered to be the fastest querying engine within the Hadoop distribution vendors market, and it is a direct competitor of Spark from UC Berkeley.

- **Cloudera Manager:** This is Cloudera's console to manage and deploy Hadoop components within your Hadoop cluster.

- **Hue:** A console that lets the user interact with the data and run scripts for the different Hadoop components contained in the cluster.

Figure 1-1 illustrates Cloudera's Hadoop distribution with the following component classification:

- The components in orange are part of Hadoop core stack.

- The components in pink are part of the Hadoop ecosystem project.

- The components in blue are Cloudera-specific components.

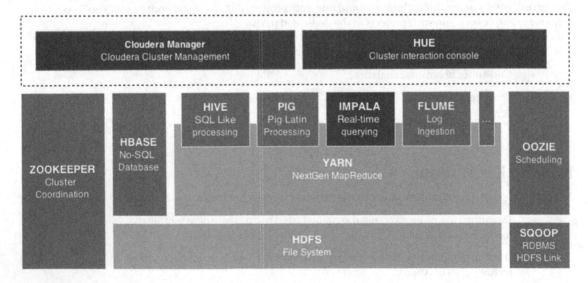

Figure 1-1. The Cloudera Hadoop distribution

Hortonworks HDP

Hortonworks is 100-percent open source and is used to package stable components rather than the last version of the Hadoop project in its distribution.

It adds a component management console to the stack that is comparable to Cloudera Manager.

Figure 1-2 shows a Hortonworks distribution with the same classification that appeared in Figure 1-1; the difference is that the components in green are Hortonworks-specific components.

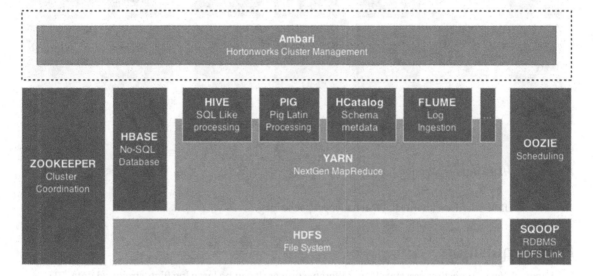

Figure 1-2. *Hortonworks Hadoop distribution*

As I said before, these two distributions (Hortonworks and Cloudera) are equivalent when it comes to building our architecture. Nevertheless, if we consider the maturity of each distribution, then the one we should choose is Cloudera; the Cloudera Manager is more complete and stable than Ambari in terms of features. Moreover, if you are considering letting the user interact in real-time with large data sets, you should definitely go with Cloudera because its performance is excellent and already proven.

Hadoop Distributed File System (HDFS)

You may be wondering where the data is stored when it is ingested into the Hadoop cluster. Generally it ends up in a dedicated file system called HDFS.

These are HDFS's key features:

- Distribution

- High-throughput access

- High availability

- Fault tolerance

- Tuning

- Security

- Load balancing

HDFS is the first class citizen for data storage in a Hadoop cluster. Data is automatically replicated across the cluster data nodes.

Figure 1-3 shows how the data in HDFS can be replicated over a cluster of five nodes.

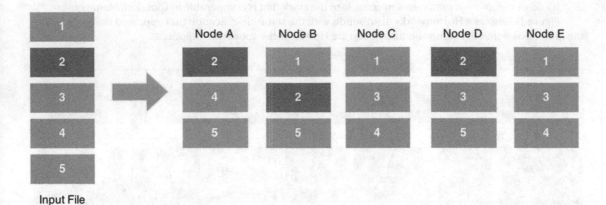

Figure 1-3. *HDFS data replication*

You can find out more about HDFS at hadoop.apache.org.

Data Acquisition

Data acquisition or ingestion can start from different sources. It can be large log files, streamed data, ETL processing outcome, online unstructured data, or offline structure data.

Apache Flume

When you are looking to produce ingesting logs, I would highly recommend that you use Apache Flume; it's designed to be reliable and highly available and it provides a simple, flexible, and intuitive programming model based on streaming data flows. Basically, you can configure a data pipeline without a single line of code, only through configuration.

Flume is composed of sources, channels, and sinks. The Flume source basically consumes an event from an external source, such as an Apache Avro source, and stores it into the channel. The channel is a passive storage system like a file system; it holds the event until a sink consumes it. The sink consumes the event, deletes it from the channel, and distributes it to an external target.

Figure 1-4 describes the log flow between a web server, such as Apache, and HDFS through a Flume pipeline.

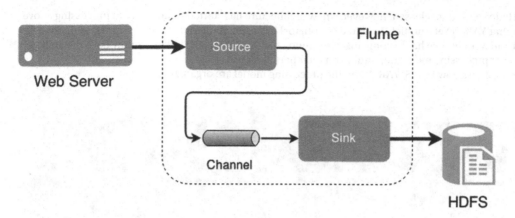

Figure 1-4. Flume architecture

With Flume, the idea is to use it to move different log files that are generated by the web servers to HDFS, for example. Remember that we are likely to work on a distributed architecture that might have load balancers, HTTP servers, application servers, access logs, and so on. We can leverage all these assets in different ways and they can be handled by a Flume pipeline. You can find out more about Flume at flume.apache.org.

Apache Sqoop

Sqoop is a project designed to transfer bulk data between a structured data store and HDFS. You can use it to either import data from an external relational database to HDFS, Hive, or even HBase, or to export data from your Hadoop cluster to a relational database or data warehouse.

Sqoop supports major relational databases such as Oracle, MySQL, and Postgres. This project saves you from writing scripts to transfer the data; instead, it provides you with performance data transfers features.

Since the data can grow quickly in our relational database, it's better to identity fast growing tables from the beginning and use Sqoop to periodically transfer the data in Hadoop so it can be analyzed.

Then, from the moment the data is in Hadoop, it is combined with other data, and at the end, we can use Sqoop export to inject the data in our business intelligence (BI) analytics tools. You can find out more about Sqoop at sqoop.apache.org.

Processing Language

Once the data is in HDFS, we use a different processing language to get the best of our raw bulk data.

Yarn: NextGen MapReduce

MapReduce was the main processing framework in the first generation of the Hadoop cluster; it basically grouped sibling data together (Map) and then aggregated the data in depending on a specified aggregation operation (Reduce).

In Hadoop 1.0, users had the option of writing MapReduce jobs in different languages—Java, Python, Pig, Hive, and so on. Whatever the users chose as a language, everyone relied on the same processing model: MapReduce.

Since Hadoop 2.0 was released, however, a new architecture has started handling data processing above HDFS. Now that YARN (Yet Another Resource Negotiator) has been implemented, others processing models are allowed and MapReduce has become just one among them. This means that users now have the ability to use a specific processing model depending on their particular use case.

Figure 1-5 shows how HDFS, YARN, and the processing model are organized.

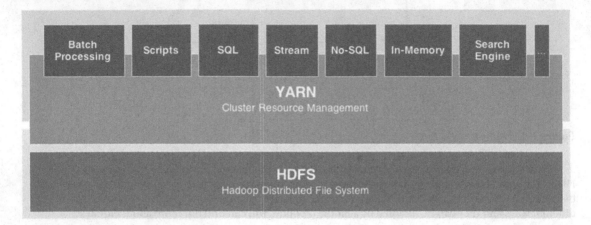

Figure 1-5. *YARN structure*

We can't afford to see all the language and processing models; instead we'll focus on Hive and Spark, which cover our use cases, namely long data processing and streaming.

Batch Processing with Hive

When you decide to write your first batch-processing job, you can implement it using your preferred programming language, such as Java or Python, but if you do, you better be really comfortable with the mapping and reducing design pattern, which requires development time and complex coding, and is, sometimes, really hard to maintain.

As an alternative, you can use a higher-level language, such as Hive, which brings users the simplicity and power of querying data from HDFS in a SQL-like way. Whereas you sometimes need 10 lines of code in MapReduce/Java; in Hive, you will need just one simple SQL query.

When you use another language rather than using native MapReduce, the main drawback is the performance. There is a natural latency between Hive and MapReduce; in addition, the performance of the user SQL query can be really different from a query to another one, as is the case in a relational database. You can find out more about Hive at hive.apache.org.

Hive is not a near or real-time processing language; it's used for batch processing such as a long-term processing job with a low priority. To process data as it comes, we need to use Spark Streaming.

Stream Processing with Spark Streaming

Spark Streaming lets you write a processing job as you would do for batch processing in Java, Scale, or Python, but for processing data as you stream it. This can be really appropriate when you deal with high throughput data sources such as a social network (Twitter), clickstream logs, or web access logs.

Spark Streaming is an extension of Spark, which leverages its distributed data processing framework and treats streaming computation as a series of nondeterministic, micro-batch computations on small intervals. You can find out more about Spark Streaming at spark.apache.org.

Spark Streaming can get its data from a variety of sources but when it is combined, for example, with Apache Kafka, Spark Streaming can be the foundation of a strong fault-tolerant and high-performance system.

Message-Oriented Middleware with Apache Kafka

Apache is a distributed publish-subscribe messaging application written by LinkedIn in Scale. Kafka is often compared to Apache ActiveMQ or RabbitMQ, but the fundamental difference is that Kafka does not implement JMS (Java Message Service). However, Kafka is a persistent messaging and high-throughput system, it supports both queue and topic semantics, and it uses ZooKeeper to form the cluster nodes.

Kafka implements the publish-subscribe enterprise integration pattern and supports parallelism and enterprise features for performance and improved fault tolerance.

Figure 1-6 gives high-level points of view of a typical publish-subscribe architecture with message transmitting over a broker, which serves a partitioned topic.

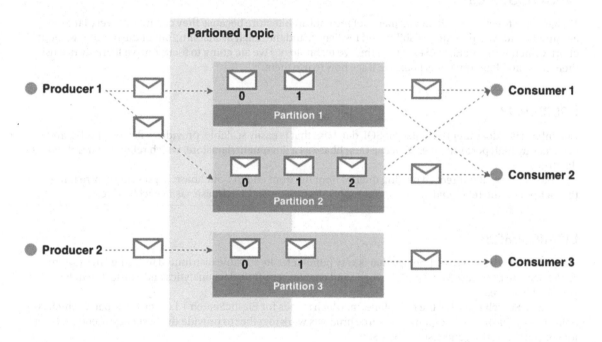

Figure 1-6. *Kafka partitioned topic example*

We'll use Kafka as a pivot point in our architecture mainly to receive data and push it into Spark Streaming. You can find out more about Kafka at kafka.apache.org.

Machine Learning

It's never too soon to talk about machine learning in our architecture, specifically when we are dealing with use cases that have an infinity converging model that can be highlighted with a small data sample. We can use a machine learning—specific language or leverage the existing layers, such as Spark with Spark MLlib (machine learning library).

Spark MLlib

MLlib enables machine learning for Spark, it leverages the Spark Direct Acyclic Graph (DAG) execution engine, and it brings a set of APIs that ease machine learning integration for Spark. It's composed of various algorithms that go from basic statistics, logistic regression, k-means clustering, and Gaussian mixtures to singular value decomposition and multinomial naive Bayes.

With Spark MLlib out-of-box algorithms, you can simply train your data and build prediction models with a few lines of code. You can learn more about Spark MLlib at `spark.apache.org/mllib`.

NoSQL Stores

NoSQL datastores are fundamental pieces of the data architecture because they can ingest a very large amount of data and provide scalability and resiliency, and thus high availability, out of the box and without effort. Couchbase and ElasticSearch are the two technologies we are going to focus on; we'll briefly discuss them now, and later on in this book, we'll see how to use them.

Couchbase

Couchbase is a document-oriented NoSQL database that is easily scalable, provides a flexible model, and is consistently high performance. We'll use Couchbase as a document datastore, which relies on our relational database.

Basically, we'll redirect all reading queries from the front end to Couchbase to prevent high-reading throughput on the relational database. For more information on Couchbase, visit `couchbase.com`.

ElasticSearch

ElasticSearch is a NoSQL technology that is very popular for its scalable distributed indexing engine and search features. It's based on Apache Lucene and enables real-time data analytics and full-text search in your architecture.

ElasticSearch is part of the ELK platform, which stands for ElasticSearch + Logstash + Kibana, which is delivered by Elastic the company. The three products work together to provide the best end-to-end platform for collecting, storing, and visualizing data:

- **Logstash** lets you collect data from many kinds of sources—such as social data, logs, messages queues, or sensors—it then supports data enrichment and transformation, and finally it transports them to an indexation system such as ElasticSearch.

- **ElasticSearch** indexes the data in a distributed, scalable, and resilient system. It's schemaless and provides libraries for multiple languages so they can easily and fatly enable real-time search and analytics in your application.

- **Kibana** is a customizable user interface in which you can build a simple to complex dashboard to explore and visualize data indexed by ElasticSearch.

Figure 1-7 shows the structure of Elastic products.

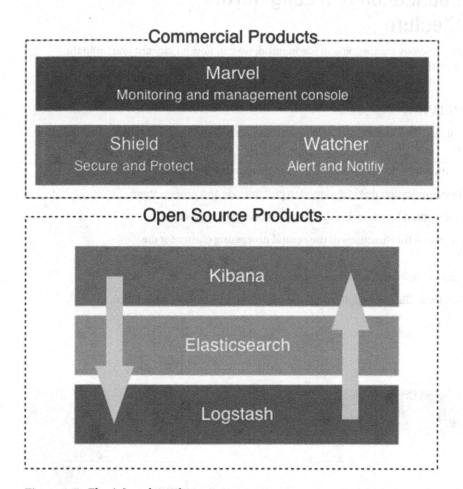

Figure 1-7. *ElasticSearch products*

As you can see in the previous diagram, Elastic also provides commercial products such as Marvel, a monitoring console based on Kibana; Shield, a security framework, which, for example, provides authentication and authorization; and Watcher, an alerting and notification system. We won't use these commercial products in this book.

Instead, we'll mainly use ElasticSearch as a search engine that holds the data produced by Spark. After being processed and aggregated, the data is indexed into ElasticSearch to enable a third-party system to query the data through the ElasticSearch querying engine. On the other side, we also use ELK for the processing logs and visualizing analytics, but from a platform operational point of view. For more information on ElasticSearch, visit `elastic.co`.

Creating the Foundation of a Long-Term Big Data Architecture

Keeping all the Big Data technology we are going to use in mind, we can now go forward and build the foundation of our architecture.

Architecture Overview

From a high-level point of view, our architecture will look like another e-commerce application architecture. We will need the following:

- A web application the visitor can use to navigate in a catalog of products

- A log ingestion application that is designed to pull the logs and process them

- A learning application for triggering recommendations for our visitor

- A processing engine that functions as the central processing cluster for the architecture

- A search engine to pull analytics for our process data

Figure 1-8 shows how these different applications are organized in such an architecture.

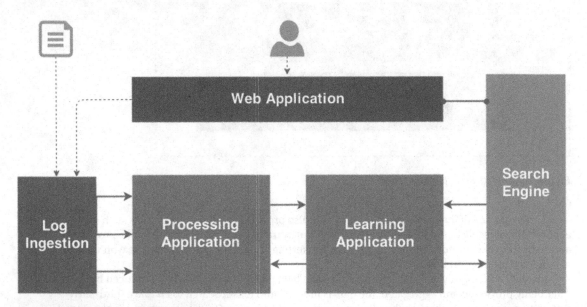

Figure 1-8. *Architecture overview*

Log Ingestion Application

The log ingestion application is used to consume application logs such as web access logs. To ease the use case, a generated web access log is provided and it simulates the behavior of visitors browsing the product catalog. These logs represent the clickstream logs that are used for long-term processing but also for real-time recommendation.

There can be two options in the architecture: the first can be ensured by Flume and can transport the logs as they come in to our processing application; the second can be ensured by ElasticSearch, Logstash, and Kibana (the ELK platform) to create access analytics.

Figure 1-9 shows how the logs are handled by ELK and Flume.

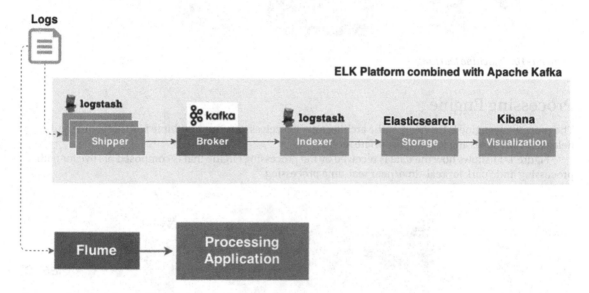

Figure 1-9. Ingestion application

Using ELK for this architecture gives us a greater value since the three products integrate seamlessly with each other and bring more value that just using Flume alone and trying to obtain the same level of features.

Learning Application

The learning application receives a stream of data and builds prediction to optimize our recommendation engine. This application uses a basic algorithm to introduce the concept of machine learning based on Spark MLlib.

Figure 1-10 shows how the data is received by the learning application in Kafka, is then sent to Spark to be processed, and finally is indexed into ElasticSearch for further usage.

Figure 1-10. *Machine learning*

Processing Engine

The processing engine is the heart of the architecture; it receives data from multiple kinds of source and delegates the processing to the appropriate model.

Figure 1-11 shows how the data is received by the processing engine that is composed of Hive for path processing and Spark for real-time/near real-time processing.

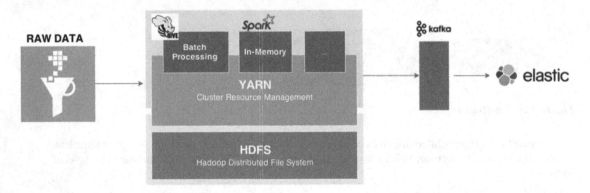

Figure 1-11. *Processing engine*

Here we use Kafka combined with Logstash to distribute the data to ElasticSearch. Spark lives on top of a Hadoop cluster, which is not mandatory. In this book, for simplicity's sake, we do not set up a Hadoop cluster, but prefer to run Spark in a standalone mode. Obviously, however, you're able to deploy your work in your preferred Hadoop distribution.

Search Engine

The search engine leverages the data processed by the processing engine and exposes a dedicated RESTful API that will be used for analytic purposes.

Summary

So far, we have seen all the components that make up our architecture. In the next chapter, we will focus on the NoSQL part and will further explore two different technologies: Couchbase and ElasticSearch.

CHAPTER 2

■ ■ .■

Early Big Data with NoSQL

In this chapter, I provide you with an overview of the available datastore technologies that are use in a Big Data project context. I then focus on Couchbase and ElasticSearch and show you how they can be used and what their differences are.

The first section gives you a better understanding of the different flavors of existing technologies within the NoSQL landscape.

NoSQL Landscape

Relational databases were the choice, almost the only choice, of a lot of developers and database administrators for traditional three-tier applications. This was the case for many reasons having to do with the data modeling methodology, the querying language that interacted with the data, and the powerful nature of those technologies, which allowed for consistent data stores to be deployed that served complex applications.

Then the needs started to evolve/change in such a way that those data stores could no longer be the solution to all data-store problems. That's how the term *NoSQL* arose—it offered a new approach to those problems by first breaking the standardized SQL schema-oriented paradigms.

NoSQL technologies are schemaless and highly scalable, and couple of them are also highly distributed and high-performance. Most of the time, they complete an architecture with an existing RDBMS technology by, for example, playing the role of cache, search engine, unstructured store, and volatile information store.

They are divided in four main categories:

1. Key/value data store

2. Column data store

3. Document-oriented data store

4. Graph data store

Now let's dive into the different categories and then choose the most appropriate for our use case.

Key/Value

The first and easiest NoSQL data stores to understand are key/value data stores. These data stores basically act like a dictionary and work by matching a key to a value. They are often used for high-performance use cases in which basic information needs to be stored—for example, when session information may need to be written and retrieved very quickly. These data stores really perform well and are efficient for this kind of use case; they are also usually highly scalable.

Key/value data stores can also be used in a queuing context to ensure that data won't be lost, such as in logging architecture or search engine indexing architecture use cases.

Redis and Riak KV are the most famous key/value data stores; Redis is more widely used and has an in-memory K/V store with optional persistence. Redis is often used in web applications to store session-related data, like node or PHP web applications; it can serve thousands of session retrievals per second without altering the performance. Another typical use case is the queuing use case that I describe later in this book; Redis is positioned between Logstash and ElasticSearch to avoid losing streamed log data before it is indexed in ElasticSearch for querying.

Column

Column-oriented data stores are used when key/value data stores reach their limits because you want to store a very large number of records with a very large amount of information that goes beyond the simple nature of the key/value store.

Column data store technologies might be difficult to understand for people coming from the RDBMS world, but actually, they are quite simple. Whereas data is stored in rows in RDBMS, it is obviously stored in columns in column data stores. The main benefit of using columnar databases is that you can quickly access a large amount of data. Whereas a row in an RDBMS is a continuous disk entry, and multiple rows are stored in different disk locations, which makes them more difficult to access, in columnar databases, all cells that are part of a column are stored continuously.

As an example, consider performing a lookup for all blog titles in an RDBMS; it might be costly in terms of disk entries, specifically if we are talking about millions of records, whereas in columnar databases, such a search would represent only one access.

Such databases are indeed very handy for retrieving large amounts of data from a specific family, but the tradeoff is that they lack flexibility. The most used columnar database is Google Cloud Bigtable, but specifically, Apache HBase and Cassandra.

One of the other benefits of columnar databases is ease of scaling because data is stored in columns; these columns are highly scalable in terms of the amount of information they can store. This is why they are mainly used for keeping nonvolatile, long-living information and in scaling use cases.

Document

Columnar databases are not the best for structuring data that contains deeper nesting structures—that's where document-oriented data stores come into play. Data is indeed stored into key/value pairs, but these are all compressed into what is called a *document*. This document relies on a structure or encoding such as XML, but most of the time, it relies on JSON (JavaScript Object Notation).

Although document-oriented databases are more useful structurally and for representing data, they also have their downside—specifically when it comes to interacting with data. They basically need to acquire the whole document—for example, when they are reading for a specific field—and this can dramatically affect performance.

You are apt to use document-oriented databases when you need to nest information. For instance, think of how you would represent an account in your application. It would have the following:

- **Basic information:** first name, last name, birthday, profile picture, URL, creation date, and so on

- **Complex information:** address, authentication method (password, Facebook, etc.), interests, and so on

That's also why NoSQL document-oriented stores are so often used in web applications: representing an object with nested object is pretty easy, and integrating with front-end JavaScript technology is seamless because both technologies work with JSON.

The most used technologies today are MongoDB, Couchbase, and Apache CouchDB. These are easy to install and start, are well documented, and are scalable, but above all, they are the most obvious choices for starting a modern web application.

Couchbase is one the technologies we are going to use in our architecture specifically because of the way we can store, organize, and query the data using it. I made the choice of Couchbase mainly based on a performance benchmark that reveals that high latency is lower for high operation thoughputs than it is in MongoDB.

Also it's worth mentioning that Couchbase is the combination of CouchDB and Memcached, and today, from a support perspective, it makes more sense to use Couchbase, more details on this link.

Graph

Graph databases are really different from other types of database. They use a different paradigm to represent the data—a tree-like structure with nodes and edges that are connected to each other through paths called *relations*.

Those databases rise with social networks to, for example, represent a user's friends network, their friends' relationships, and so on. With the other types of data stores, it is possible to store a friend's relationship to a user in a document, but still, it can then be really complex to store friends' relationships; it's better to use a graph database and create nodes for each friend, connect them through relationships, and browse the graph depending on the need and scope of the query.

The most famous graph database is Neo4j, and as I mentioned before, it is used for use cases that have to do with complex relationship information, such as connections between entities and others entities that are related to them; but it is also used in classification use cases.

Figure 2-1 shows how three entities would be connected within a graph database.

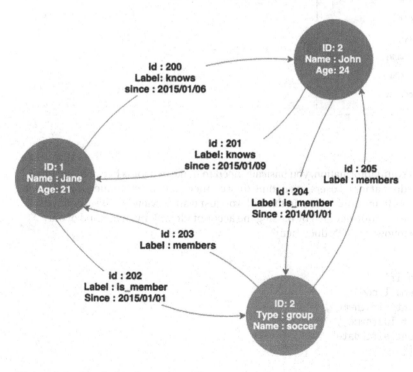

Figure 2-1. *Graph database example*

The diagram's two accounts nodes, Jane and John, connect to each other through edges that define their relationship; they have known each other since a defined date. Another group node connects to the two accounts nodes and this shows that Jane and John have been part of the soccer group since a defined date.

NoSQL in Our Use Case

In the context of our use case, we first need a document-oriented NoSQL technology that structures the data contained in our relational database into a JSON document. As mentioned earlier, traditional RDBMSs store data into multiple tables linked with relationships, which makes it harder and less efficient when you want to get the description of a whole object.

Let's take the example of an account that can be split into the tables shown in Figure 2-2.

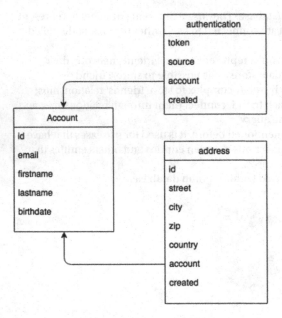

Figure 2-2. *Account tables*

If you want to retrieve all account information, you basically need to make two joins between the three tables. Now think this: I need to do that for all users, every time they connect, and these connections happen for different business logic reasons in my application. In the end, you just want a "view" of the account itself.

What if we can get the whole account view just by passing the account identifier to a method of our application API that returns the following JSON document?

```
{

    "id": "account_identifier",
    "email": "account@email.com",
    "firstname": "account_firstname",
    "lastname": "account_lastname",
    "birthdate": "account_birthdate",
    "authentication": [{
```

```
                "token": "authentication_token_1",
                "source": "authenticaton_source_1",
                "created": "12-12-12"
        }, {
                "token": "authentication_token_2",
                "source": "authenticaton_source_2",
                "created": "12-12-12"
        }],
        "address": [{
                "street": "address_street_1",
                "city": "address_city_1"
                "zip": "address_zip_1"
                "country": "address_country_1"
                "created": "12-12-12"
        }]
}
```

The benefit is obvious: by keeping a fresh JSON representation of our application entities, we can get better and faster data access. Furthermore, we can generalize this approach to all read requests so that all the reading is done on the NoSQL data store, which leaves all the modification requests (create, update, delete) to be made on the RDBMS. But we must implements a logic that transactionally spreads any changes on the RDBMS to the NoSQL data store and also that creates the object from the relational database if it is not found in the cache.

You may wonder why we would keep the RDBMS when we know that creating documents in a NoSQL data store is really efficient and scalable. It is because that is actually not the goal of our application. We don't want to make a Big Bang effect. Let's assume that the RDBMS was already there and that we want to integrate a NoSQL data store because of the lack of flexibility in a RDBMS. We want to leverage the best of the two technologies—specifically the data consistency in the RDBMS and the scalability from the NoSQL side.

Beside, this is just a simple query example that we can perform, but we want to go further by, for example, making full-text searches on any field of our document. Indeed, how do we do this with a relational database? There is indexing, that's true, but would we index all table columns? In fact, that's not possible; but this is something you can easily do with NoSQL technologies such as ElasticSearch.

Before we dive into such a NoSQL caching system, we need to go through how to use a Couchbase document-oriented database, and then we need to review the limitations that will drive us to switch to ElasticSearch.

We will see that our scalable architecture first relies on Couchbase, but because of some important Couchbase limitations, we'll first complete the architecture with ElasticSearch before we make a definitive shift to it.

Introducing Couchbase

Couchbase is an open source, document-oriented database that has a flexible data model, is performant, is scalable, and is suitable for applications like the one in our use case that needs to shift its relational database data into a structured JSON document.

Most NoSQL technologies have similar architectures—we'll first see how the Couchbase architecture is organized and get introduced to naming convention in Couchbase, then we'll go deeper into detail on how querying data is stored in Couchbase, and finally we'll talk about cross datacenter replication.

Architecture

Couchbase is based on a real shared-nothing architecture, which means that there is no single point of contention because every node in the cluster is self-sufficient and independent. That's how distributed technologies work—nodes don't share any memory or disk storage.

Documents are stored in JSON or in binary in Couchbase, are replicated over the cluster, and are organized into units called *buckets*. A bucket can be scaled depending on the storage and access needs by setting the RAM for caching and also by setting the number of replication for resiliency. Under the hood, a bucket is split into smaller units called *vBuckets* that are actually data partitions. Couchbase uses a cluster map to map the partition to the server to which it belongs.

A Couchbase server replicates up to three times a bucket within a cluster; every Couchbase server then manages a subset of the active or replica vBuckets. That's how resiliency works in Couchbase; every time a document is indexed, it's replicated, and if a node within the cluster goes down, then the cluster promotes a replica partition to active to ensure continuous service.

Only one copy of the data is active with zero or more replicas in the cluster as Figure 2-3 illustrates.

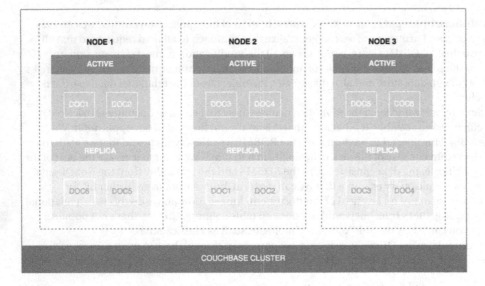

Figure 2-3. *Couchbase active document and replicas*

From a client point of view, if smart-clients are used as part as the provided clients (Java, C, C++, Ruby, etc.), then these clients are connected to the cluster map; that's how clients can send requests from applications to the appropriate server, which holds the document. In term of interaction, there is an important point to remember: operations on documents are, by default, asynchronous. This means that when, for example, you update a document, Couchbase does not update it immediately on the disk. It actually goes through the processing shown in Figure 2-4.

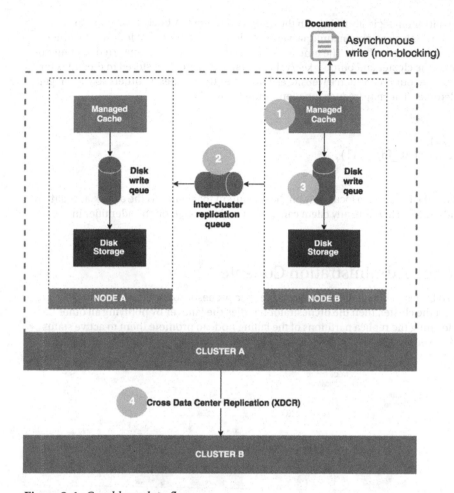

Figure 2-4. *Couchbase data flow*

As Figure 2-4 shows, the smart-client connects to a Couchbase server instance and first asynchronously writes the document in the managed cache. The client gets a response immediately and is not blocked until the end of the data flow process, but this behavior can be changed at the client level to make the client wait for the write to be finished. Then the document is put in the inter-cluster write queue, so the document is replicated across the cluster; after that, the document is put in the disk storage write queue to be persisted on the related node disk. If multiple clusters are deployed, then the Cross Data Center Replication (XDCR) feature can be used to propagate the changes to other clusters, located on a different data center.

Couchbase has its own way to query the data; indeed, you can query the data with a simple document ID, but the power of Couchbase is inside the view feature. In Couchbase, there is a second-level index called the design document, which is created within a bucket. A bucket can contain multiple types of document, for example, in a simple e-commerce application a bucket would contain the following:

- Account

- Product

- Cart

- Orders

- Bills

The way Couchbase splits them logically is through the design document. A bucket can contain multiple design documents, which also contain multiple views. A view is a function that indexes documents contained in the bucket in a user-defined way. The function is precisely a user-defined map/reduce function that maps documents across the cluster and outputs key/value pairs, which are then stored in the index for further retrieval. Let's go back to our e-commerce website example and try to index all orders so we can get them from the account identifier. The map/reduce function would be as follows:

```
function(doc, meta) {
if (doc.order_account_id)
        emit(doc.order_account_id, null);
}
```

The if statement allows the function to focus only on the document that contains the order_account_id field and then index this identifier. Therefore any client can query the data based on this identifier in Couchbase.

Cluster Manager and Administration Console

Cluster manager is handled by a specific node within the cluster, the orchestrator node. At any time, if one of the nodes fails within the cluster, then the orchestrator handles the failover by notifying all other nodes within the cluster, locating the replica partitions of the failing node to promote them to active status. Figure 2-5 describes the failover process.

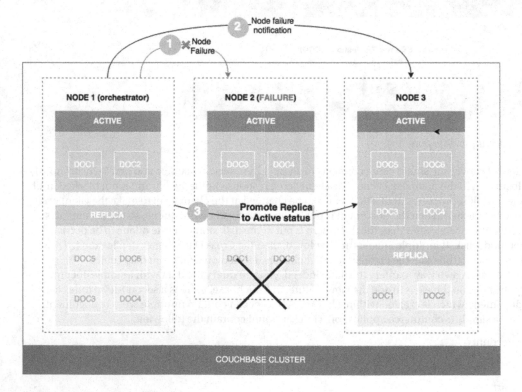

Figure 2-5. *Couchbase failover*

If the orchestrator node fails, then all nodes detect that through the heartbeat watchdog, which is a cluster component that runs on all cluster nodes. Once the failure is detected, a new orchestrator is elected among the nodes.

All cluster-related features are exposed through APIs that can be used to manage Couchbase, but the good news is that an administration console is shipped out of the box. Couchbase console is a secure console that lets you manage and monitor your cluster; you can choose from the available actions, which include setting up your server, creating buckets, browsing and updating documents, implementing new views, and monitoring vBucket and the disk write queue.

Figure 2-6 shows the Couchbase console home page with an overview of the RAM used by existing buckets, the disk used by data, and the buckets' activity.

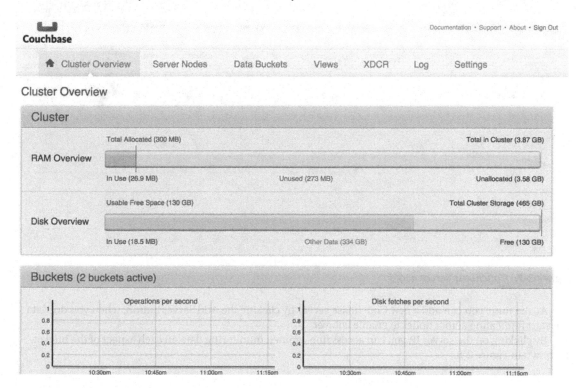

Figure 2-6. *Couchbase console home*

You can perform cluster management in the Server Nodes tab, which lets the user configure failover and replication to prevent them from losing data. Figure 2-7 shows a single node installation that is not safe for failover as the warning mentions.

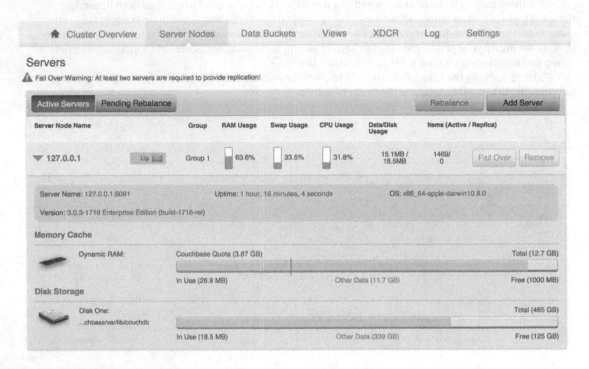

Figure 2-7. *Couchbase server nodes*

At any time, you can add a new Couchbase server by clicking the Add Server button; when you do, data will start replicating across nodes to enable failover.

By clicking on the server IP, you can access fine-grained monitoring data on each aspect of the bucket, as shown in Figure 2-8.

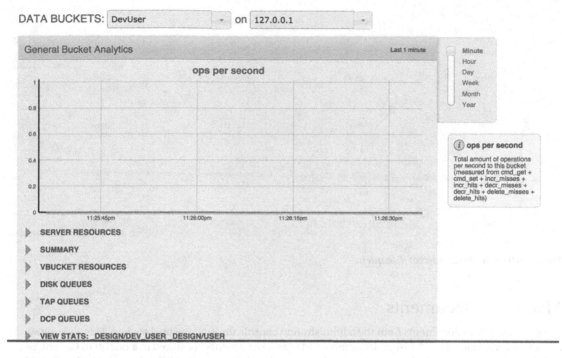

Figure 2-8. *Couchbase bucket monitoring*

This figure shows a data bucket called DevUser that contains the user-related JSON document. As explained earlier, the process of indexing a new document is part of a complex data flow under the hood. The metrics shown in the monitoring console are essential when you are dealing with a large amount of data that generates a high indexing throughput. For example, the disk queue statistics can reveal bottlenecks when data is being written on the disk.

In Figure 2-9, we can see that the *drain rate*—the number of items written on the disk from the disk write queue—is alternatively flat on the active side when the replica is written, and that the average age of the active item grows during that flat period. An altering behavior would have been to see the average age of the active item keep growing, which would mean that the writing process was too slow compared to the number of active items pushed into the write disk queue.

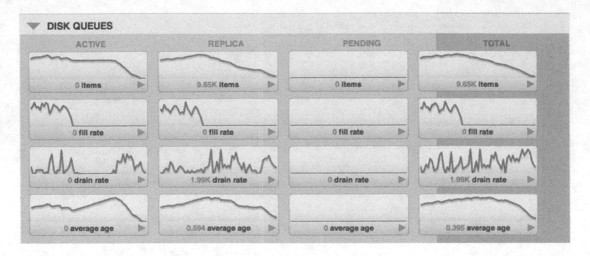

Figure 2-9. *Couchbase bucket disk queue*

Managing Documents

You can manage all documents from the administration console through the bucket view. This view allows users to browse buckets and design documents and views. Documents are stored in a bucket in Couchbase, and they can be accessed in the Data Bucket tab on the administration console as shown in the Figure 2-10.

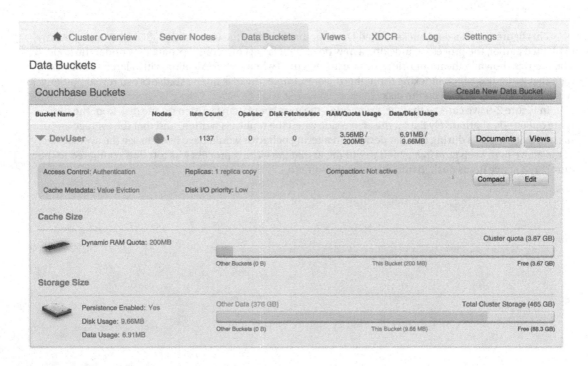

Figure 2-10. *Couchbase console bucket view*

As in the server view, the console gives statistics on the bucket, such as RAM and storage size, as well as the number of operation per second. But the real benefit of this view is that you are able to browse documents and retrieve them by ID as is shown in Figure 2-11.

Figure 2-11. Couchbase document by ID

It's also in this view that you create a design document and views to index documents for further retrieval, as shown in Figure 2-12.

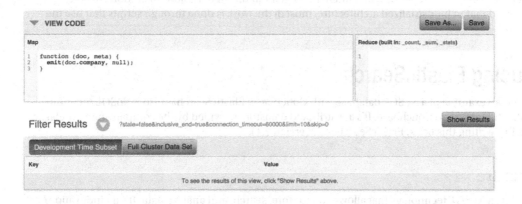

Figure 2-12. Couchbase console view implementation

In Figure 2-12, I have implemented a view that retrieves documents based on the company name. The administration console is a handy way to manage documents, but in real life, you can start implementing your design document in the administration console, and you can create a backup to industrialize its deployment.

All design documents are stored in a JSON file and a simple structure that describes all the views, similar to what Listing 2-1 shows.

Listing 2-1. Designing a Document JSON Example

```
[{
    ...
    "doc": {
        "json": {
            "views": {
                "by_Id": {"map": "function (doc, meta)
                {\n  emit(doc.id, doc);\n}"},
                "by_email": {"map": "function (doc, meta)
                {\n  emit(doc.email, doc);\n}"},
                "by_name": {"map": "function (doc, meta)
                {\n  emit(doc.firstname, null);\n}"},
                "by_company": {"map": "function (doc, meta)
                {\n  emit(doc.company, null);\n}"},
                "by_form": {"map": "function (doc, meta)
                {\n  emit(meta.id, null);\n}"}
            }
        }
        ...
    }
}]
```

As you have seen, you can perform document management through the administration console, but keep in mind that in industrialized architecture, most of the work is done through scripts that use the Couchbase API.

Introducing ElasticSearch

You have seen an example of a NoSQL database with Couchbase; ElasticSearch is also a NoSQL technology but it's totally different than Couchbase. It's a distributed datastore provided by the company named Elastic (at the time I'm writing this book, ElasticSearch is in version 2.1).

Architecture

ElasticSearch is a NoSQL technology that allows you to store, search, and analyze data. It's an indexation/search engine made on top of Apache Lucene, an open source full-text search engine written in Java. From the start, ElasticSearch was made to be distributed and to scale out, which means that in addition to scaling ElasticSearch vertically by adding more resource to a node, you can simply scale it horizontally by adding more nodes on the fly to increase the high availability of your cluster but also its resiliency. In the case of a node failure, because data is replicated over the cluster, data is served by another node.

30

ElasticSearch is a schemaless engine; data is stored in JSON and is partitioned into what we call *shards*. A shard is actually a Lucene index and is the smallest unit of scale in ElasticSearch. Shards are organized in *indexes* in ElasticSearch with which an application can make read and write interactions. In the end, an index is just a logical namespace in ElasticSearch that regroups a collection of shards, and when a request comes in, ElasticSearch routes it to the appropriate shards as the Figure 2-13 summarizes.

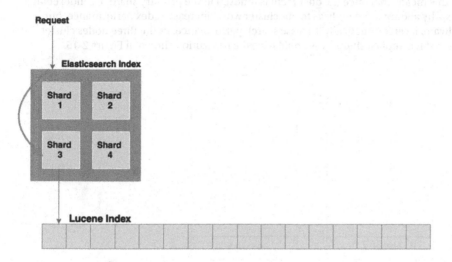

Figure 2-13. *Elasticsearch index and shards*

Two types of shards live in ElasticSearch: primary shards and replica shards. When you start an ElasticSearch node, you can begin by adding only one primary shard, which might be enough, but what if the read/index request throughput increases with time? If this is the case, the one primary shard might not be enough anymore and you then need another shard. You can't add shards on the fly and expect ElasticSearch to scale; it will have to re-index all data in the bigger index with the two new primary shards. So, as you can see, from the beginning of a project based on ElasticSearch, it's important that you have a decent estimate of how many primary shards you need in the cluster. Adding more shards in a node may not increase the capacity of your cluster, because you are limited to the node hardware capacity. To increase cluster capacity, you have to add more nodes that hold primary shards as well, as is shown in Figure 2-14.

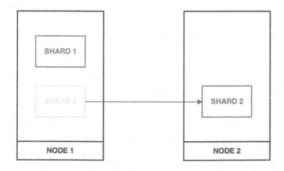

Figure 2-14. *ElasticSearch primary shard allocation*

The good thing is that ElasticSearch automatically copies the shard over the network on the new node for you as described in Figure 2-14. But what if you want to be sure that you won't lose data? That's where replica shards come into play.

Replica shards are made at start for failover; when a primary shard dies, a replica is promoted to become the primary to ensure continuity in the cluster. Replica shards have the same load that primary shards do at index time; this means that once the document is indexed in the primary shard, it's indexed in the replica shards. That's why adding more replicas to our cluster won't increase index performance, but still, if we add extra hardware, it can dramatically increase search performance. In the three nodes cluster, with two primary shards and two replica shards, we would have the repartition shown in Figure 2-15.

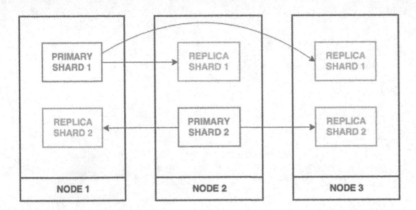

Figure 2-15. *ElasticSearch primary and replica shards*

In addition to increased performance, this use case also helps with balancing requests and getting better performance on the overall cluster.

The last thing to talk about in terms of pure ElasticSearch architecture is indices, and more specifically, nodes. Indices are regrouped into ElasticSearch nodes, and there are actually three types of nodes as shown in Figure 2-16.

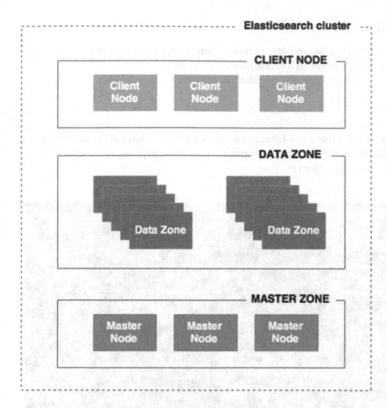

Figure 2-16. *ElasticSearch cluster topology*

Here are descriptions of the three types of node:

- **Master nodes:** These nodes are lightweight and responsible for cluster management. It means they don't hold any data, server indices, or search requests. They are dedicated to ensure cluster stability and have a low workload. It's recommended that you have three dedicated master nodes to be sure that if one fails, redundancy will be guaranteed.

- **Data nodes:** These nodes hold the data and serve index and search requests.

- **Client nodes:** These ensure load balancing in some processing steps and can take on part of the workload that a data node can perform, such as in a search request where scattering the request over the nodes and gathering the output for responses can be really heavy at runtime.

Now that :you understand the architecture of ElasticSearch, let's play with the search API and run some queries.

Monitoring ElasticSearch

Elastic provides a plug-in called Marvel for ElasticSearch that aims to monitor an ElasticSearch cluster. This plug-in is part of Elastic's commercial offer, but you can use it for free in Development mode.

You can download Marvel from the following link; the installation process is quite simple:

```
https://www.elastic.co/downloads/marvel
```

Marvel relies on Kibana, the visualization console of Elastic, and comes with a bunch of visualization techniques that let an operator be really precise about what happens in the cluster. Figure 2-17 shows the overview dashboard that you see when launching Marvel.

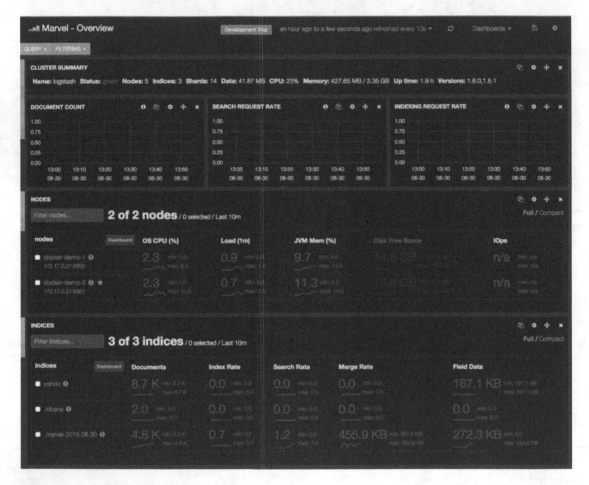

Figure 2-17. *ElasticSearch Marvel console*

Marvel provides information on nodes, indices, and shards; about the CPU used; about the memory used by the JVM; about the indexation rate, and about the search rate. Marvel even goes down to the Lucene level by providing information about flushes and merges. You can, for example, have a live view of the shard allocation on the cluster, as shown in Figure 2-18.

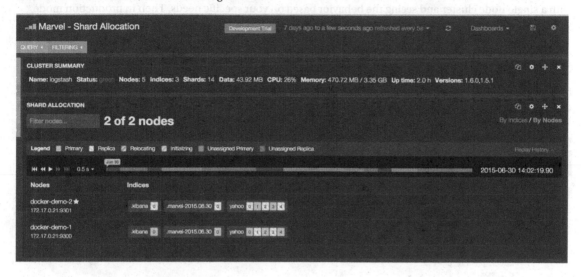

Figure 2-18. *Marvel's shard allocation view*

To give you an idea of the amount of information that Marvel can provide you with about your cluster, Figure 2-19 shows a subset of what you get in the Node Statistics dashboard.

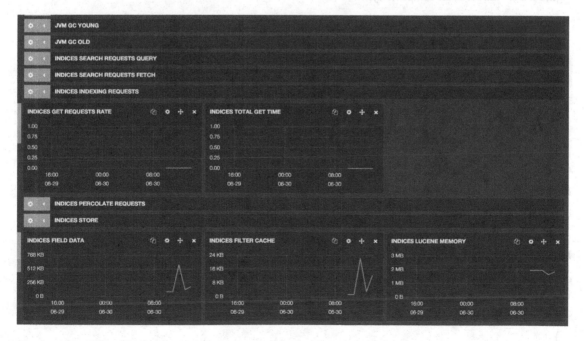

Figure 2-19. *Marvel Node Statistics dashboard*

As you can see, the dashboard is organized in several rows; in fact, there are more than 20 rows that you just can't see in this screenshot. Each row contains one or visualization about the row topic. In Figure 2-19, you can see that no GET requests are sent to indices; that's why the line chart is flat on 0.

During the development mode, these statistics will help you scale your server by, for example, starting with a single node cluster and seeing the behavior based on your specific needs. Then in production mode, you are able to follow the life inside your cluster without losing any information about it.

Search with ElasticSearch

Marvel comes with a feature called Sense that is a query editor/helper for ElasticSearch. The real power of Sense is its ability to autocomplete your query, which helps a lot when you are not familiar with all ElasticSearch APIs, as shown in Figure 2-20.

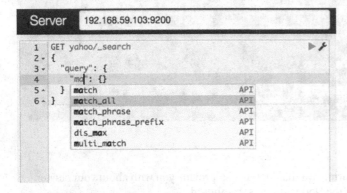

Figure 2-20. Marvel Sense completion feature

You can also export your query to cURLs, for example, so you can use it in a script as show in Figure 2-21.

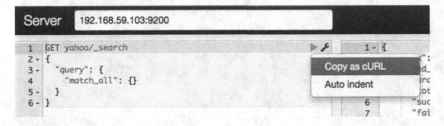

Figure 2-21. Marvel Sense copy as cURL feature

In this case, the query would give a cURL command, as show in Listing 2-3.

Listing 2-3. Example of cURL Command Generated by Sense

```
curl -XGET "http://192.168.59.103:9200/yahoo/_search" -d'
{
  "query": {
    "match_all": {}
  }
}'
```

This query basically searches for all documents under the Yahoo index. To demonstrate the benefit of the search API, I'm using a dataset from Yahoo that contains the prices of Yahoo stock since couple of years. One of the key features in ElasticSearch's search API is the aggregation framework. You can aggregate data in a different way; the most common way is through a date histogram, which is equivalent to a timeline. An example of the query is illustrated in Figure 2-22; this aggregates the stock data by date, with an interval that corresponds to a month and also calculates the max value of the stock for the given month.

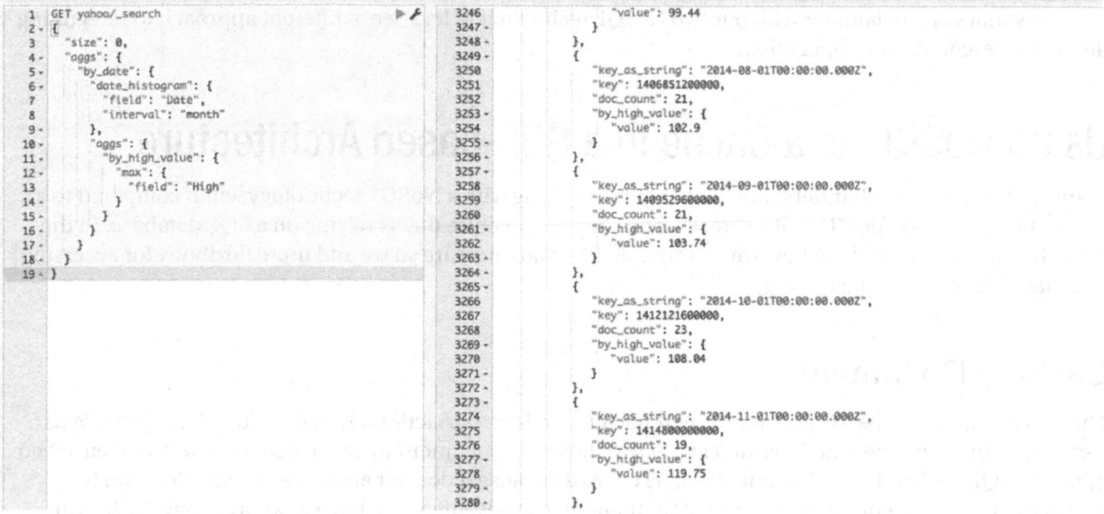

Figure 2-22. *Marvel Sense Aggregation example*

As a result, we get a document with an array where each item contains the date, the number of the document found in the month period, and the highest value, as shown in Listing 2-4.

Listing 2-4. Aggregation Bucket

```
{
        "key_as_string": "2015-01-01T00:00:00.000Z",
        "key": 1420070400000,
        "doc_count": 20,
        "by_high_value": {
                "value": 120
        }
}
```

As you can see in the query in Figure 2-22, there are two levels of aggregation: the first is for the date histogram, and the second is for the max value. Indeed, ElasticSearch supports multiple levels of aggregation as long as it make sense from a query point of view.

The search API is really rich, and we can't browse all features one by one in this book, but you get an introduction to its capabilities and get more information about it on the following documentation link:

```
https://www.elastic.co/guide/en/elasticsearch/reference/1.4/search-search.html
```

Now that you are familiar with our two NoSQL technologies, let's see a different approach to integrating them in an e-commerce application.

Using NoSQL as a Cache in a SQL-based Architecture

At this point, you should understand the benefit of working with a NoSQL technology when compared to a SQL database. But we don't want to break an existing architecture that is relying on a SQL database. In the following approach, we'll see how we can complete our architecture so we add more flexibility for accessing data based on a NoSQL technology.

Caching Document

The first thing we need to discuss is how to replicate our data transactionally in our NoSQL backend. What we want is for each time data is written in a SQL database, a document to be created, updated, and enriched in our NoSQL backend. The document can be created because it doesn't already exist, or a field can be updated, or it can be enriched with a subdocument that corresponds to a table relation in our RDBMS. In terms of accessing the document, whenever an API GET request is made, the underlying implementation should first look in the NoSQL backend and return the document if it's there.

But what if the document is not present? Then a cache miss event should be triggered and sent to the NoSQL manager implementation to rebuild the document from RDBMS. Sounds great! But what if the insert transaction fails at the SQL level? The framework should be transactional and only trigger a document build if the transaction has been committed in SQL.

Figure 2-23 summarizes the mechanism.

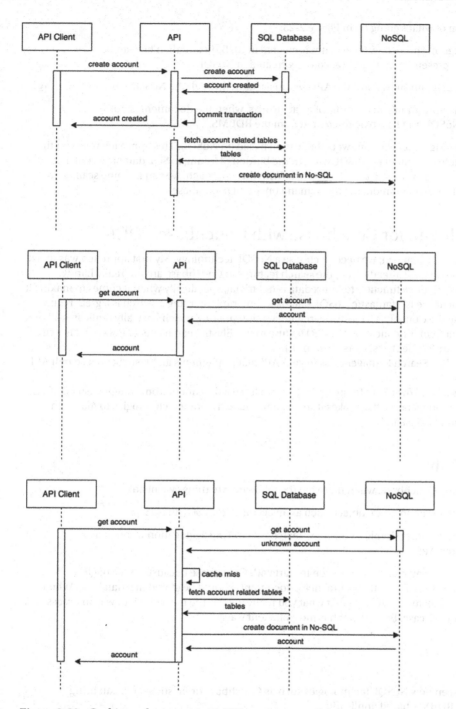

Figure 2-23. *Caching a document in NoSQL*

Here is a description of what's going on in Figure 2-23:

- The first diagram shows how an account is created in RDBMS and then how a document representing the whole account is created in NoSQL.

- The second diagram shows how the API gets an account through the NoSQL store.

- The third diagram shows an example of a cache miss, when the document is not present in NoSQL and has to be re-created from the RDBMS.

We are actually building a cache for our web application based on NoSQL. This approach relies on the flexibility to access a whole document in NoSQL without the burden of asking to SQL database, and also the flexibility to leverage the distributed aspect of NoSQL. With this approach, we can imagine scaling our cluster as the request rate grows and reducing the demand on our SQL database.

ElasticSearch Plug-in for Couchbase with Couchbase XDCR

To achieve such a caching mechanism, we need to choose a NoSQL technology. My first approach was to use Couchbase alone, but as you have already seen, the search features in Couchbase are not really handy.

Dealing with the map/reduce function to index data is not trivial, specifically when it comes to making a simple aggregation. Couchbase is a fantastic NoSQL database, but implementing a search is a great pain.

Instead, you can use the Couchbase/ElasticSearch integration plug-in, which basically replicates all the data using the Cross Data Center Replication (XDCR) feature to our ElasticSearch cluster (`www.couchbase.com/jp/couchbase-server/connectors/elasticsearch`).

This way, you can add a Search Manager class to your API that implements an ElasticSearch search API to retrieve the data.

But still, this causes a lot of overhead for just being able to fetch a document from NoSQL instead of our RDBMS. In terms of operational tasks, this makes it so you have three different technologies to maintain: RDBMS, Couchbase, and ElasticSearch.

ElasticSearch Only

Why should we keep using Couchbase, when these are the following are things we need?

- The ability to index an entire object, such as an account from SQL to NoSQL

- The ability to conduct simple to complex searches involving aggregation through a flexible search API.

In our specific use case, you could have chosen to start with Couchbase because it's a NoSQL database and, as you know from learned best practices, documents should be stored in a database. When experiencing such an architecture, you learn that what you really need is the most efficient way to access and request your data, and in our case, ElasticSearch is most efficient way.

Summary

At this point, you have seen how NoSQL technologies such as Couchbase or ElasticSearch can bring flexibility to an existing RDBMS-based application.

Also, you have seen that two technologies that are perceived as similar from a high-level point of view can be fundamentally different.

In the next chapter, you learn how to deal with a processing topology.

CHAPTER 3

■ ■ ■

Defining the Processing Topology

In the previous chapter, I explained how to use a NoSQL technology in the architecture that you're building in this book. The NoSQL technology provides some caching and searching features to the application, but since you want to process data, you need to define an approach to handle various streams of data so you can deliver insights to your users or data services to your applications.

This chapter will explain how to define the processing topology by walking you through a data architecture that is commonly used in IT organizations. Then you'll see why splitting is something that comes naturally when respecting service level agreements (SLAs). Finally, I'll discuss a specific kind of architecture, the lambda architecture, which is a combination of different types of architectures.

First Approach to Data Architecture

This section describes the first step of deploying a data architecture using common technologies from the Big Data space.

A Little Bit of Background

A lot of IT organizations are evolving their data architecture today because they are relying on what you could call traditional data architectures. Dealing with multiple data sources is not new at all; these architectures are connected to a multitude of sources such as CRM systems, file systems, and legacy systems. They are mainly running on top of relational databases such as Oracle, DB2, or Microsoft SQL.

Today, the usual data analytics cycle is to run some periodic processes that extract and process data directly from the databases. That is mainly done today by extract, transform, and load (ETL) tools such as Informatica or Talend. The goal is to load the digested data into a data warehouse for further analytics.

Unfortunately, this way of handling data isn't inline with the business needs at the end of the cycle. These data pipelines can take hours, days, and sometimes weeks to complete, but business decisions need to be made now.

Besides the processing time, there are some fundamental changes in the nature of data that can't be handled easily by these architectures, such as a data structure that forces the data modeling to be refactored or even a data volume that leads to scalability considerations. Scaling those systems isn't easy because I'm not talking about a distributed system here. Scaling is expensive in terms of deployment time but also in terms of hardware because databases require highly performant CPUs, RAM, and storage solutions to run.

Today most IT organizations are switching to deploying data architectures based on Hadoop technologies. Indeed, rather than relying on inflexible, costly technology, the objective is now to distribute the load of processing over a bunch of commodity machines and be able to ingest a large amount of different types of data.

Figure 3-1 gives an overview of the topology of such an architecture.

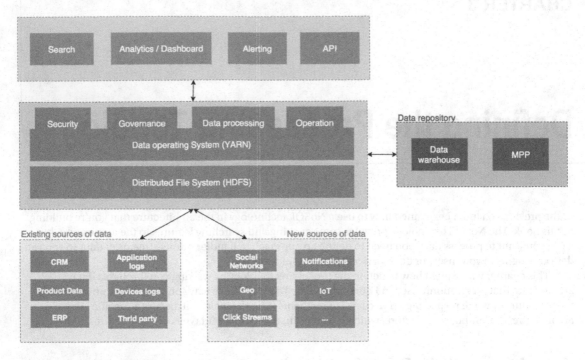

Figure 3-1. *Hadoop-based data architecture*

The following section what covers the data pipeline, what kind of technologies are involved, and the common practices for this type of architecture.

Dealing with the Data Sources

As Figure 3-1 shows, the data can come from multiple internal or external sources, but Big Data particularly comes from internal application and device logs, as well as from social networks, open data, or even sensors.

If you take the example of a social network, data is flooding into IT organizations with interesting information, but this attractive information is drowning among a massive amount of useless information. Therefore, first this data needs to be stored, and then it needs to be processed to extract the important parts. This data is really useful for sales, specifically when you run sentiment analysis on top of it to reveal your social ecosystem's feeling about your brand or products.

Depending on the provider, data can be structured, semistructured, or unstructured. Listing 3-1 gives an example of a semistructured message.

Listing 3-1. Example of the Semistructured Data of a Tweet

```
{
    "created_at": "Fri Sep 11 12:11:59 +0000 2015",
    "id": 642309610196598800,
    "id_str": "642309610196598785",
    "text": "After a period in silent mode, going back to tweet life",
    "source": "<a href="http://twitter.com/download/iphone" rel="nofollow">
    Twitter for iPhone</a>",
    "truncated": false,
```

```
"in_reply_to_status_id": null,
"in_reply_to_status_id_str": null,
"in_reply_to_user_id": null,
"in_reply_to_user_id_str": null,
"in_reply_to_screen_name": null,
"user": {
  "id": 19450096,
  "id_str": "19450096",
  "name": "Bahaaldine",
  "screen_name": "Bahaaldine",
  "location": "Paris",
  "description": "",
  "url": null,
  "entities": {
    "description": {
      "urls": []
    }
  },
  "protected": false,
  "followers_count": 59,
  "friends_count": 107,
  "listed_count": 8,
  "created_at": "Sat Jan 24 15:32:11 +0000 2009",
  "favourites_count": 66,
  "utc_offset": null,
  "time_zone": null,
  "geo_enabled": true,
  "verified": false,
  "statuses_count": 253,
  "lang": "en",
  "contributors_enabled": false,
  "is_translator": false,
  "is_translation_enabled": false,
  "profile_background_color": "C0DEED",
  "profile_background_image_url": "http://pbs.twimg.com/profile_background_
images/454627542842896384/-n_C_Vzs.jpeg",
  "profile_background_image_url_https": "https://pbs.twimg.com/profile_background_
images/454627542842896384/-n_C_Vzs.jpeg",
  "profile_background_tile": false,
  "profile_image_url": "http://pbs.twimg.com/profile_images/448905079673094144/
dz109X55_normal.jpeg",
  "profile_image_url_https": "https://pbs.twimg.com/profile_images/448905079673094144/
dz109X55_normal.jpeg",
  "profile_banner_url": "https://pbs.twimg.com/profile_banners/19450096/1397226440",
  "profile_link_color": "0084B4",
  "profile_sidebar_border_color": "FFFFFF",
  "profile_sidebar_fill_color": "DDEEF6",
  "profile_text_color": "333333",
  "profile_use_background_image": true,
  "has_extended_profile": false,
  "default_profile": false,
```

```
      "default_profile_image": false,
      "following": false,
      "follow_request_sent": false,
      "notifications": false
    },
    "geo": null,
    "coordinates": null,
    "place": null,
    "contributors": null,
    "is_quote_status": false,
    "retweet_count": 0,
    "favorite_count": 0,
    "entities": {
      "hashtags": [],
      "symbols": [],
      "user_mentions": [],
      "urls": []
    },
    "favorited": false,
    "retweeted": false,
    "lang": "en"
  }
```

As you can see in the example, the document is a JavaScript Object Notation (JSON) document that has a bunch of fields for the string metadata that describes the tweet. But some of the fields are more complex than that; there are arrays that are sometimes empty and sometimes containing a collection of data. There is also plain text such as the tweet content.

This should lead you to think about how your data needs to be stored. Putting your data on HDFS isn't enough at all; you will likely have to build a metastructure on top of it with technologies that support such complexity in data structures. This is something you can achieve with Hive, for example, and is discussed in the following sections.

Social networks are representative of the complexity you get when you deal with a large amount of heterogeneous data. Besides the structure, you need to classify the data into logical subsets in order to enhance the data processing.

If you think about the example of sentiment analysis, it makes sense to organize the data to enable the location of valuable information in large sets of unstructured data. As an example, a common practice is to partition the data into time-based partitions so further data processing can easily focus, for example, on a specific week amound years of data.

You also have to take care of securing the access to the data, most likely by using a Kerberos or another authentication provider. But as the data platform attracts new use cases, the first thing to handle with multitenancy is security. Then you need to create periodic snapshots of your data to archive it and retrieve it in case of failure.

All these considerations are pretty standard and fortunately offered by most distribution vendors. There are out-of-the-box packages that ensure or will help you to implement or configure these concepts.

Processing the Data

Data transformations are often handled by ETL tools when you want a pure transformation from a source to a target. Tools such as Talend, Pentaho, Informatica, or IBM Datastage are the most commonly used software in Big Data projects. But still, they are not self-sufficient, and you will need some complementary tools (such Sqoop) for simply importing or exporting the data. In any case, multiple tools are used to ingest the data, and what they have in common is the target: HDFS. HDFS is the entry point in Hadoop distribution; data needs somehow to be stored in this file system in order to be processed by high-level applications and projects.

When the data is stored on HDFS, then how do you access and process it?

This what Hive, for example, can do. It creates a structure on top of the files stored in HDFS to make it easier to access the files. The structure itself is like a table in a data. For example, Listing 3-2 shows a structure that can be created for the preceding tweet example.

Listing 3-2. Hive Tweet Structure

```
create table tweets (
   created_at string,
   entities struct <
      hashtags: array ,
            text: string>>,
      media: array ,
            media_url: string,
            media_url_https: string,
            sizes: array >,
            url: string>>,
      urls: array ,
            url: string>>,
      user_mentions: array ,
            name: string,
            screen_name: string>>>,
   geo struct <
      coordinates: array ,
      type: string>,
   id bigint,
   id_str string,
   in_reply_to_screen_name string,
   in_reply_to_status_id bigint,
   in_reply_to_status_id_str string,
   in_reply_to_user_id int,
   in_reply_to_user_id_str string,
   retweeted_status struct <
      created_at: string,
      entities: struct <
         hashtags: array ,
               text: string>>,
         media: array ,
               media_url: string,
               media_url_https: string,
               sizes: array >,
               url: string>>,
         urls: array ,
               url: string>>,
         user_mentions: array ,
```

```
                name: string,
                screen_name: string>>>,
        geo: struct <
            coordinates: array ,
            type: string>,
        id: bigint,
        id_str: string,
        in_reply_to_screen_name: string,
        in_reply_to_status_id: bigint,
        in_reply_to_status_id_str: string,
        in_reply_to_user_id: int,
        in_reply_to_user_id_str: string,
        source: string,
        text: string,
        user: struct <
            id: int,
            id_str: string,
            name: string,
            profile_image_url_https: string,
            protected: boolean,
            screen_name: string,
            verified: boolean>>,
    source string,
    text string,
    user struct <
        id: int,
        id_str: binary,
        name: string,
        profile_image_url_https: string,
        protected: boolean,
        screen_name: string,
        verified: boolean>
)
PARTITIONED BY (datehour INT)
ROW FORMAT SERDE 'org.openx.data.jsonserde.JsonSerDe'
LOCATION '/user/username/tweets';
```

As you can see, the tweets are structures in a table, with a substructure describing the complexity of the source's unstructured document. Now the data is securely stored in HDFS, structured with Hive, and ready to be part of a processing or querying pipeline. As an example, Listing 3-3 shows the distribution of hashtags within the collected data.

Listing 3-3. Top Hashtags

```
SELECT
 LOWER(hashtags.text),
 COUNT(*) AS hashtag_count
FROM tweets
LATERAL VIEW EXPLODE(entities.hashtags) t1 AS hashtags
GROUP BY LOWER(hashtags.text)
ORDER BY hashtag_count DESC
LIMIT 15;
```

Querying data with Hive is really convenient because it provides a SQL-like querying language. The problem is the latency of the query; it basically has the same latency of a MapReduce job. Indeed, the previous query is then translated into a MapReduce job that follows the usual execution pipeline, which results in long-running processing.

This situation becomes problematic when the need is to follow data transformations in real time, such as when watching top hashtags in live. Real-time massive data processing isn't a myth anymore with the rise of technology such as Spark. It brings real-time XX but also is simple to implement. For example, Listing 3-4 shows how you would implement a word count in MapReduce.

Listing 3-4. Word Count in MapReduce (from www.dattamsha.com/2014/09/ hadoop-mr-vs-spark-rdd-wordcount-program/)

```
package org.apache.hadoop.examples;

import java.io.IOException;
import java.util.StringTokenizer;

import org.apache.hadoop.conf.Configuration;
import org.apache.hadoop.fs.Path;
import org.apache.hadoop.io.IntWritable;
import org.apache.hadoop.io.Text;
import org.apache.hadoop.mapreduce.Job;
import org.apache.hadoop.mapreduce.Mapper;
import org.apache.hadoop.mapreduce.Reducer;
import org.apache.hadoop.mapreduce.lib.input.FileInputFormat;
import org.apache.hadoop.mapreduce.lib.input.FileSplit;
import org.apache.hadoop.mapreduce.lib.output.FileOutputFormat;
import org.apache.hadoop.util.GenericOptionsParser;

public class WordCount {

    public static class TokenizerMapper extends
        Mapper<Object, Text, Text, IntWritable> {

        private final static IntWritable one = new IntWritable(1);
        private Text word = new Text();

        public void map(Object key, Text value, Context context)
            throws IOException, InterruptedException {
            StringTokenizer itr = new StringTokenizer(value.toString());
            while (itr.hasMoreTokens()) {
                word.set(itr.nextToken());
                context.write(word, one);
            }
        }
    }

    public static class IntSumReducer extends
        Reducer<Text, IntWritable, Text, IntWritable> {

        private IntWritable result = new IntWritable();
```

47

```java
        public void reduce(Text key, Iterable<IntWritable> values,
        Context context) throws IOException, InterruptedException {
            int sum = 0;
            for (IntWritable val : values) {
                sum += val.get();
            }
            result.set(sum);
            context.write(key, result);
        }
    }

    public static void main(String[] args) throws Exception {
        Configuration conf = new Configuration();
        String[] otherArgs = new GenericOptionsParser(conf, args)
            .getRemainingArgs();

        Job job = new Job(conf, "word count");

        job.setJarByClass(WordCount.class);

        job.setMapperClass(TokenizerMapper.class);
        job.setCombinerClass(IntSumReducer.class);
        job.setReducerClass(IntSumReducer.class);

        job.setOutputKeyClass(Text.class);
        job.setOutputValueClass(IntWritable.class);

        FileInputFormat.addInputPath(job, new Path(otherArgs[0]));
        FileOutputFormat.setOutputPath(job, new Path(otherArgs[1]));

        System.exit(job.waitForCompletion(true) ? 0 : 1);
    }
}
```

Listing 3-5 shows how you would do this in Spark (Python).

Listing 3-5. Word Count in Spark

```python
from pyspark import SparkContext
logFile = "hdfs://localhost:9000/user/bigdatavm/input"
sc = SparkContext("spark://bigdata-vm:7077", "WordCount")
textFile = sc.textFile(logFile)
wordCounts = textFile.flatMap(lambda line: line.split()).map(lambda word: (word, 1)).
reduceByKey(lambda a, b: a+b)
wordCounts.saveAsTextFile("hdfs://localhost:9000/user/bigdatavm/output")
```

That leads me to discuss the need to split the architecture into multiple parts that handle specific needs, one for the batch processing and the other one for the streaming.

Splitting the Architecture

Using Hadoop brings a lot of solutions when it comes to handling a large amount of data, but it also brings challenges when allocating resources and managing the stored data; you always want to reduce the cost while keeping minimum latency to provide data to end users.

Like in any other architecture, data architecture should answer to the needs driven by the SLAs. Thus, every job should not consume every resource equally and should be either managed via a prioritization system or isolated from each other in terms of architecture, hardware, network, and so on.

In this section, I'll talk about how modern data architecture is split to address the different levels of time needed because of SLAs. To illustrate, I'll discuss the repartition illustrated in Figure 3-2.

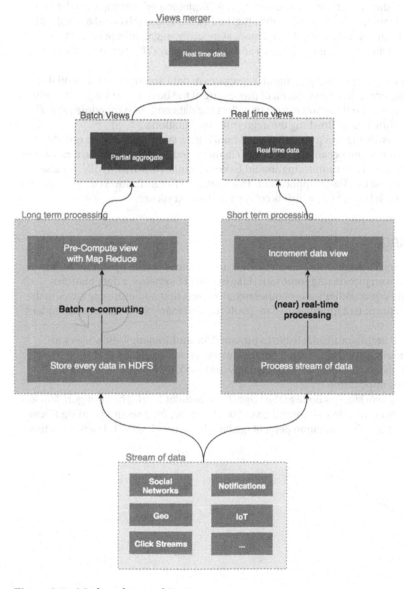

Figure 3-2. *Modern data architecture*

As you can see, the architecture is divided into the following:

- A long-term processing part
- A short-term processing part
- A view merging part

I'll go through each of these parts and explain what role is being affected.

Batch Processing

The long-term processing jobs, or batch processing, are jobs that were implemented starting with the first generation of Hadoop architecture, such as MapReduce, Hive, Pig, and so on. Those jobs tend to deal with massive amounts of data and come up with digested data or, more standardly, aggregates of data. Data is distributed using HDFS file system capabilities and schedules and is governed by different tools depending on the distribution you use.

Generally, the objective of these jobs is to keep compute aggregates of data and make them available for analytics. As said, batch processing was a first-class citizen of processing at the beginning of Big Data trends because it was a natural way to process the data: you extract or collect the data and then schedule jobs. The processing jobs can spend a lot of time before finishing the aggregate computations.

These jobs were also mainly answering to a processing need from legacy systems that were not able to deal with the data stream. Batch processing became really easy to manage and monitor because it ran in one shot, compared to a streaming system where monitoring should be continuous. Now with YARN, it's also easy to manage the resources allocated to a batch application. This way, the IT organization can also split the batch architecture depending on each batch SLA, which is covered in the next section.

Prioritizing the Processing

When dealing with batch processing, IT organizations want to have total control on the operations around processing, such as scheduling or even prioritizing some jobs. Like in most IT systems, a data platform starts on a pilot use case that might attract other parts of other organizations that add more use cases to the platform. This simply turns the platform to a multitenant data platform with many SLAs that are dependent on use cases.

In Hadoop 2.0 and Yarn-based architecture, the features provided for multitenancy allow users to access the same data platform and run processing with different cluster capacities. YARN also allows you to run non-MapReduce applications, so with the ResourceManager and the YARN Capacity Scheduler, prioritizing jobs can be done across application types.

The distribution of the Hadoop workload is done at the Capacity Scheduler level. The configuration lets you finely arrange a predictable cluster resource share and gives you the ability for a secured and significant cluster utilization. This works by setting the utilization percentage on job queues. Figure 3-3 illustrates this concept.

Yarn Capacity Scheduler

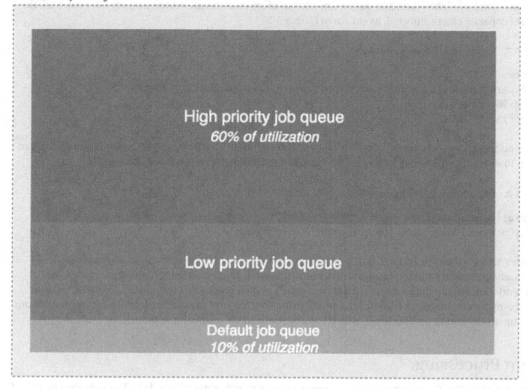

Figure 3-3. *YARN job queue*

The example illustrates different priorities for the three queues: high, low, and default. This can be translated in the simple YARN capacity scheduler configuration shown in Listing 3-6.

Listing 3-6. YARN Capacity Scheduler Configuration

```
<property>
  <name>yarn.scheduler.capacity.root.queues</name>
  <value>default,highPriority,lowPriority</value>
</property>
<property>
  <name>yarn.scheduler.capacity.root.highPriority.capacity</name>
  <value>60</value>
</property>
<property>
  <name>yarn.scheduler.capacity.root.lowPriority.capacity</name>
  <value>20</value>
</property>
<property>
  <name>yarn.scheduler.capacity.root.default.capacity</name>
  <value>10</value>
</property>
```

Each queue has been affected a minimum cluster capacity, which by the way is elastic. This means if idle resources are available, then one queue can take up more than the minimum initially affected. Still, the maximum capacity can be affected, as shown in Listing 3-7.

Listing 3-7. Maximum Queue Capacity

```
<property>
  <name>yarn.scheduler.capacity.root.lowPriority.maximum-capacity</name>
  <value>50</value>
</property>
```

This configuration sets the capacity, so anyone who submits a job (for example, a MapReduce job), can submit it to a specific queue depending on the expected requirements, as shown in Listing 3-8.

Listing 3-8. Submitting a MapReduce Job

```
Configuration priorityConf = new Configuration();
priorityConf.set("mapreduce.job.queuename", queueName);
```

With YARN Capacity Scheduler, batch processing becomes even more efficient in terms of resource management and has a lot of applications in the industry, such as allocating more resources for the recommendation engine than for less critical needs such as data processing for notifications. But talking about recommendations, most IT systems now are using shorter-term processing jobs for recommendations and relying on streaming architectures, which I will describe in the following section.

Stream Processing

Short-term processing, or stream processing, intends to ingest high-throughput data. The solutions designed for streaming can handle a high volume of data and are highly distributed, scalable, and fault-tolerant technologies.

This kind of architecture has to resolve a bunch of challenges. As said, the main one is to process massive amounts of data. There were existing streaming technologies before, but this one should be highly available, resilient, and performant. They are performant because data volume, complexity, and size increase.

If data volume increases, that's because these architectures have to be able to integrate with a multitude of data sources and applications seamlessly, such as data warehouses, files, data history, social networks, application logs, and so on. They need to offer a consistent agile API and client-oriented API and also be able to output information on various channels such as notification engines, search engines, and third-party applications. Basically such technologies have much more constraints in terms of real-time responsiveness to market changes.

Finally, the capacity of delivering real-time analytics is the first expectation end users have from a streaming architecture. The needs are pretty clear and consist of the following: discovering data in real time, being able to query data as easily as possible, and proactive monitoring through alerting to notify users and applications whenever an event reaches a threshold.

Streaming architecture was first used in the financial sector with a high-throughput trading use case, but it has expanded to a various number of use cases, mainly e-commerce, telecommunications, fraud detection, and analytics. This has caused the rise of two main technologies: Apache Spark and Apache Storm.

Spark is the technology I've chosen to rely on in this book. I've had the chance to work with it a couple of times, and I think it's the one that has the most traction, value, and support from the community. Figure 3-4 supports this.

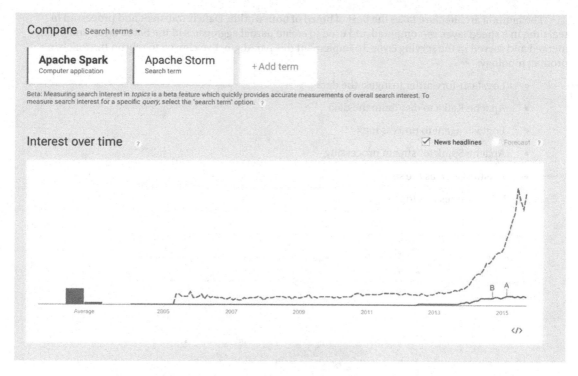

Figure 3-4. *Google trends of Apache Spark and Apache Storm*

Rather than going into more detail in this section, I'm dedicating two chapters to the streaming and serving architecture: Chapters X and X. I'll describe how to combine different technologies including Spark to handle live streams and search analytics.

The Concept of a Lambda Architecture

The previous sections described the need for splitting a data architecture into three parts: batching, streaming, and serving the architecture. While batch processing is still a common practice in most existing IT organizations' data architecture, it doesn't justify a real need when you can for most of the use case first stream the data and, if it's needed, store the data in the batch-oriented file system.

Deploying a streaming architecture is a lot easier to say than to do when the IT organization's existing practices are batch oriented. Compared to it, a streaming architecture brings much more complexity in terms of operations because operators always have to be on and architects have to size and design a solution that can absorb unexpected bursts of data while maintaining a decent response time.

Still, starting an approach by streaming is a lot of easier at the end of the day when you realize the burden of deploying a Hadoop distribution "just" to do what a streaming architecture can do with the same processing API and sometimes more. Even if your SLAs are not severe and you don't expect to get the data in seconds or minutes, you are reducing the amount of energy you put in to reach them.

From my point of view, a streaming architecture is a natural evolution of the modern data architecture. It reduces the software complexity, like the reduction of hardware that happened when the first generation of Big Data arose. My architecture choices here might not be generic and are really tight with what I'm working on every day, but when I browse the various use cases I've worked on, I realize that my technology stack answers to 90 percent of the use cases.

The lambda architecture takes the best of breed of both worlds. Data is transient and processed in real time in a speed layer, re-computed and used to create partial aggregates in the batch layer, and finally merged and served in the serving layer. To implement this paradigm, I've chosen to rely on the following product topology:

- Logstash-forwarder to ingest the data
- Apache Kafka to distribute the data
- Logtash agent to process logs
- Apache Spark for stream processing
- Elasticsearch as the serving layer

Figure 3-5 illustrates this topology.

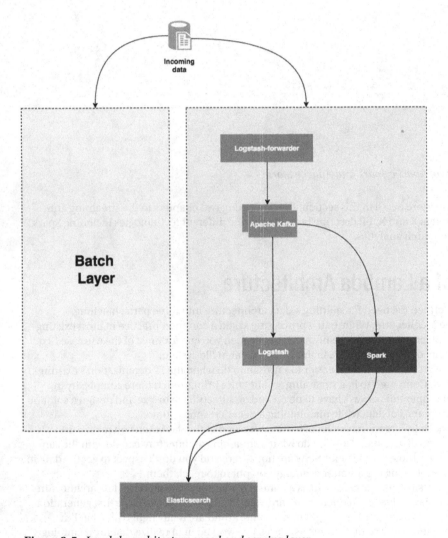

Figure 3-5. *Lambda architecture speed and serving layer*

A lambda architecture is often used for e-commerce web sites for different purposes such as recommendations or security analytics. In the case of clickstream data, you can extract multiple insights.

- On one hand, using the long-processing layer, the clickstream can be processed, aggregated, and correlated to other data sources to build an accurate source for the recommendation engine that will consume them. In this example, clickstream data can be correlated with other data lakes containing precise demographic information to build views that can be then indexed into Elasticsearch.

- On the other hand, the same data can be used for bot detection. Indeed, most e-commerce web sites experience security threats, so one way to prevent that is to analyze the click behavior through the stream layer and blacklist in real time bot IP addresses. In Figure 3-5, you can use Spark to make complex correlation or run machine-learning processes to extract data before indexing in Elasticsearch.

I won't go into the details of the batch layer in the next chapter as this is a well-known subject that most Big Data books are covering today. The serving layer is most commonly handled by Hive, but why should you use Hive while Elasticsearch is able to create analytics in near real time? Now, with new connectors such as the ES-Hadoop connector (`https://www.elastic.co/products/hadoop`), you can delegate in most use cases the data access, querying, and aggregation to real-time search technologies such as Elasticsearch, which provides even more capabilities.

Summary

At this point, you should have a better understanding about the value of the lambda architecture when it comes time to deal with a multitude of data sources types and, at the same time, about meeting the SLAs expected when it comes time to deliver services to end users and applications.

In the next chapter, I'll focus on building the stream part of the lambda architecture and will use the clickstream data to base your discussion.

CHAPTER 4

■ ■ ■

Streaming Data

In the previous chapter, we focused on a long-term processing job, which runs in a Hadoop cluster and leverages YARN or Hive. In this chapter, I would like to introduce you to what I call the 2014 way of processing the data: streaming data. Indeed, more and more data processing infrastructures are relying on streaming or logging architecture that ingest the data, make some transformation, and then transport the data to a data persistency layer.

This chapter will focus on three key technologies: Kafka, Spark, and the ELK stack from Elastic. We will work on combining them to implement different kind of logging architecture that can be used depending on the needs.

Streaming Architecture

In this section, we'll see how the streaming part of the architecture is structured and also what technology we can use for data ingestion.

Architecture Diagram

First, before going through the architecture details, let's discuss the word **streaming,** which can be misunderstand in the context of that chapter. Indeed by **streaming architecture**, I mean here an architecture that is capable to ingest data as they come.

Data can be generated by different providers, application log files such as Apache Hatted access logs, your REST API logs, from a message broker technology such as Apache Kafka, or even directly from a TCP endpoint.

Generally, the data are ingested and then transported to a buffer layer most of the time based on a message broker. We call this part the shipping part of the logging architecture. The buffer part is often used to ensure that no data will be lost and decouple the architecture between the shipping part and the processing part.

The message broker should be distributed and guarantees the persistence of data whenever data are streamed in the architecture.

After buffering, two kind of processing architecture can be set up:

- A processing pipeline that aims to transform, format, validate, and enrich the data before persisting it. Most of the time the technology that ensures that shipping handles this part. Some technologies also handle the shipping, buffering, and processing parts.

- A processing pipeline that handles Hadoop-based technology that provides extra feature to the processing such as machine learning capabilities.

Figure 4-1 gives a graphical representation of what we just discussed.

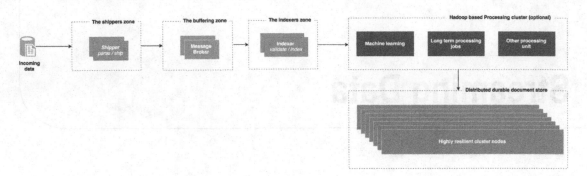

Figure 4-1. *Anatomy of a typical data ingesting architecture*

Technologies

Let's put some names now to those high-level blocks. For the ingestion part, the most used technologies might be Apache flume and Logstash from Elastic. Those are the two preferred technologies to ingest stream data from multiple varieties of data sources, process them, and then index them.

Flume architecture is scalable and relies on the three-tier structure shown in Figure 4-2.

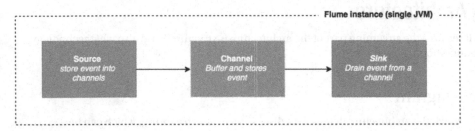

Figure 4-2. *Flume architecture*

The flume source receives data from an external source and stores it in one or more flume channel. Once in the channel, the data stored and kept until the sink properly drains it. The sink writes the message in the target store. Most of the time flume is used in Hadoop architecture because of the simplicity to write in HDFS, for example. Flume is capable to guarantee data delivery from a bunch of data sources such as Avro, Thrift, Syslog, or another flume agent; to HDFS, Log files, Avro, or directly to HBase and even Elasticsearch. Its horizontal scalability ensures fault tolerance and high throughput.

On the other side, Logstash is totally different from a structure point of view. Logstash is a single block agent that receives, forwards, and forgets the log. By forget, I don't mean that Logstash doesn't keep track of what it just consumed (a file, for example) but that it doesn't come with a buffer layer that is used to persist the data in order to provide resiliency out of the box. Logstash is written in JRuby and run within a JVM. To prevent downsides of using a JVM for transporting logs, Logstash comes with Logstash-forwarder, which are specific agents used to receive and forward the logs to another program.

The typical usage of Logstash relies on the architecture shown in Figure 4-3.

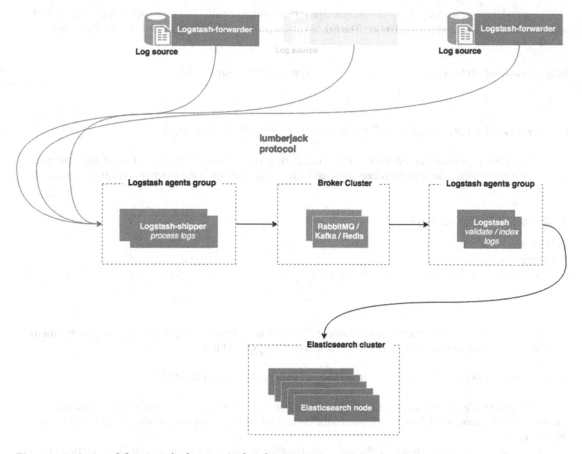

Figure 4-3. *Logstash logging platform typical architecture*

As we can see, the architecture is divided following the structure discussed earlier, with the following specificity:

- The first part is composed of Logstash-forwarders installed next to the log sources, which only receives the data and send them to a Logstash instances group responsible of processing the logs before putting them into a message broker.

- The second part is based on a message broker; the recommended ones are RabbitMQ, Kafka, and Redis. For the sake of simplicity and because it would take many pages to go through each and every possible combination, we'll focused on Apache Kafka, which tends to have the best using traction these days.

- The third part is based on the Logstash processors group, which can validate events before indexing them into Elasticsearch.

In Logstash, a processing pipeline is a declarative configuration–oriented file, which is composed of inputs, filters, and outputs.

Logstash comes with a palette of inputs/outputs that lets you connect to a variety of data sources such as file, syslog, imp, tcp, xmpp, and more. The full list of supported inputs are available on the following pages:

https://www.elastic.co/guide/en/logstash/current/input-plugins.html

and

https://www.elastic.co/guide/en/logstash/current/output-plugins.html

Filters help processing and events by only configuring a filter depending on your need. For example, if you want to convert a field of the received event to an integer, all you have to write is what's shown in Listing 4-1.

Listing 4-1. Logstash filter example

```
filter {
  mutate {
    convert => { "fieldname" => "integer" }
  }
}
```

We will examine in deeper detail an example of a processing pipeline that read logs and process them later in this chapter, but for now, here is the link to list of supported filters:

https://www.elastic.co/guide/en/logstash/current/filter-plugins.html

You might have noticed in the diagram that the proposal here is to index the data in Elasticsearch once it has been processed. Logstash supports different type of stores and has a seamless integration with Elasticsearch.

Elasticsearch will play the role of the indexation engine that we'll use later on to make searches and analytics.

The Anatomy of the Ingested Data

In this part we'll discuss the structure of the data we'll to ingenst in the streaming platform. As mentioned, our goal is to leverage the activity of our website visitor. We'll then focused on what we call clickstream data.

Clickstream Data

I don't think I need to remind you about the key data that you can get from a website in terms of visitor activity. What we are about to describe here, clickstream data, is part of it.

When you run a website, you will likely want to know how your visitors behave and how you can enhance their experience. Among the challenges for a website to grow are the following:

- Site traffic

- Unique visitor

- Conversion rate

- Pay per click traffic volume

- Time on site

- ....

Having those KPI will help you answer questions including: how are my pages accessed? When? How long do my visitors stay on this or that page? Where do people prefer to spend time? How they exit? What is the typical page-to-page navigation?

Being aware of what is happening on your website is fundamental and, hopefully, our architecture is made to generate data in logs files that will help us in this way.

Typically, the clickstream data are based on the web access log, generated by servers such as Apache Server or NGINX. Depending on your logging configuration, every time a user accesses your website and navigates through pages, for each of these actions a line is written in the access logs. Figure 4-4 gives an idea of the dimension brought by a clickstream data line.

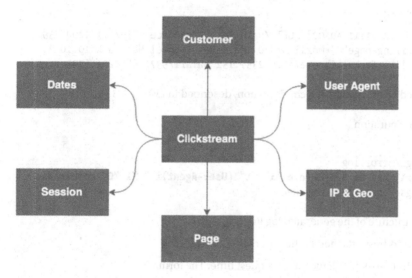

Figure 4-4. *Anatomy of a Clickstream data*

The clickstream data will give directly or indirectly:

- Dates such as the timestamp when the event occurred, the time spent by the visitor

- The user agent used to browse the website, the version, and devices

- The session attached to the visitor, which will help to make correlation between the different lines of data: the session start time, end time, duration

- The page, what the visitor browsed, what the type of request was sent to the server, the domain/subdomain

- Information about visitor IP addresses, which indirectly will give information about delocalization

- Depending on the type of information we get, it can be feasible to identity and map the customers to the clickstream data

The Raw Data

In this book, we'll work with Apache HTTP server and use the Apache logging configuration file to produce our Clickstream data. The configuration is quite simple and uses a specific syntax that makes appear in the log line information we need. Listing 4-2 is a typical access log line generate by Apache HTTP server.

Listing 4-2. Access log line example

```
10.10.10.10 - - [10/July/2015:11:1:11 +0100] "GET /path/to/my/resource HTTP/1.1" 200 4392
"http://www.domain.com/referring-page" " Mozilla/5.0 (Macintosh; Intel Mac OS X 10_10_3)
AppleWebKit/537.36 (KHTML, like Gecko) Chrome/43.0.2357.132 Safari/537.36" 1224 6793
```

This line has been produced by the Apache log configuration, described in Listing 4-3.

Listing 4-3. Apache logging configuration

```
LogLevel error
ErrorLog  /home/example/logs/error.log
LogFormat "%h %l %u %t \"%r\" %>s %b \"%{Referer}i\" \"%{User-Agent}i\" %I %O" combinedio
CustomLog /home/example/logs/access.log combinedio
```

The line in bold gives the structure of the generated log line:

- 10.10.10.10 (%h): IP address attached to the request

- [10/July/2015:11:11:11 +0100] (%t): reception request time. The format is [day/month/year:hour:minute:second zone]

- GET /path/to/my/resource HTTP/1.1 (\"%r\"): first line of the request, which contains useful information such as the request type, the path, the protocol

- 200 (%>s): the http return code

- 4392 (%b): response size return to the client without HTTP headers in bytes

- %{Referer}: the Referrer HTTP request header

- %{User-Agent}: the User Agent HTTP request header

- %I: bytes received including HTTP header

- %O: bytes sent including headers

As you can see, the data are structured by Apache in the different corresponding part but, at the end of the day, the application that will receive the logs will considered these data as unstructured, because parsing them by a third-party application is not that obvious.

You then need a technology that can consume these line as they are appended to the access log file or just ship the file, and also that is capable of transforming these data into structured document for later analysis and search.

That's where Logstash comes into play; we'll now set up our streaming architecture and start shipping and ingesting data.

The Log Generator

There might some public dataset that we would be able to use for testing our streaming architecture, but for the sake of simplicity, I'd rather implement a small log generator to create our access logs. I've chose Python for implementing that generator, as shown in Listing 4-4.

Listing 4-4. Access log code generator

```python
#!/usr/bin/python
import time
import datetime
import random
timestr = time.strftime("%Y%m%d-%H%M%S")

f = open('../source/access_log_'+timestr+'.log','w')

# ips
with open('ips.txt') as ips_file:
        ips = ips_file.read().splitlines()

# referers
with open('referers.txt') as referers_file:
        referers = referers_file.read().splitlines()

# resources
with open('resources.txt') as resources_file:
        resources = resources_file.read().splitlines()

# user agents
with open('user_agents.txt') as user_agents_file:
        useragents = user_agents_file.read().splitlines()

# codes
with open('codes.txt') as codes_file:
        codes = codes_file.read().splitlines()

# requests
with open('requests.txt') as requests_file:
        requests = requests_file.read().splitlines()

event_time = datetime.datetime(2013,10,10)
```

```
for i in xrange(0,50000):
        increment = datetime.timedelta(seconds=random.randint(30,300))
        event_time += increment
        uri = random.choice(resources)
        if uri.find("Store")>0:
                uri += `random.randint(1000,1500)`
        ip = random.choice(ips)
        useragent = random.choice(useragents)
        referer = random.choice(referers)
        code = random.choice(codes)
        request= random.choice(requests)
        f.write('%s - - [%s] "%s %s HTTP/1.0" %s %s "%s" "%s"\n' %
(random.choice(ips),event_time.strftime('%d/%b/%Y:%H:%M:%S %z'),request,uri,coe,random.randint
(2000,5000),referer,useragent))
```

This generator relies on different text files and generates in the **../source/access_log_** as much as specified when you launch the script, like in Listing 4-5.

Listing 4-5. Launching the generator

```
./generator.py 50000
```

The previous line will generate 50,000 lines of access logs. You can easily customize the content, which is used to generate the file by updating the text fles.

Note that all assets used in Listing 4-4 can be found on the following Github repo:

```
https://github.com/bahaaldine/scalable-big-data-architecture/tree/master/chapter4/generator
```

Setting Up the Streaming Architecture

To set up our streaming architecture, we'll split the work into two parts: shipping and processing.

Shipping the Logs in Apache Kafka

Shipping data consist to transport them from the machine that generates the data to the one which will ingest it, and process it. This is what we will cover here with Logstash-forwarder. Then data will be stored into an asynchronous reliable persistent layer that will ensure message delivery. The message broker here is Apache Kafka.

The Forwarders

In the following section, we will first start by installing the forwarder, and configuring it, specifically the security part ensured by Lumberjack, the secured transport protocol.

Installation

Logstash-forwarder is written in Go and can be installed either by an RPM or directly from the sources. I'll use the second way and follow these steps:

- Install Go from this URL:

  ```
  http://golang.org/doc/install
  ```

- Fetch the source from the git repository:

  ```
  git clone git://github.com/elasticsearch/logstash-forwarder.git
  ```

- And finally compile it:

  ```
  cd logstash-forwarder
  go build -o logstash-forwarder
  ```

Logstash installation is straightforward. You just need to download the appropriate package on the following page:

```
https://www.elastic.co/downloads/logstash
```

Lumberjack: The Secured Protocol

Forwarder will be installed next to the data sources, and will transport and send the logs to the Logstash processors. This uses the lumberjack protocol, which is secured and requires a valid private key (.key file) and an SSL certificate (.crt file). Both Logstash forwarder and Logstash server will use these assets to secure the data transport.

To generate a valid KEY and CRT file, run the command described in listing 4-6.

Listing 4-6. Key and Certificate command line

```
openssl req -x509  -batch -nodes -newkey rsa:2048 -keyout lumberjack.key -out
lumberjack.crt -subj /CN=localhost
```

This command will generate lumberjack.key and lumberjack.crt. An important thing to mention here is about the certificate. The subject used to generate the certificate should be the same that the one that will be used in the Logstash-forwarder file. Here we generate the certificate with a CN=local host, then in the forwarder file we'll need to use localhost as well. More information about this can be found in this page:

```
https://github.com/elastic/logstash-forwarder#important-tlsssl-certificate-notes
```

The Configuration Files

Logstash-forwarder configuration is composed of network and file subconfiguration; Listing 4-7 shows the configuration that I'm using for the example project.

Listing 4-7. Logstash configuration file, which can be found: https://github.com/bahaaldine/ scalable-big-data-architecture/blob/master/chapter4/logstash/forwarder/forwarder.json

```
{
  "network": {
    "servers": [ "HOST:PORT" ],
    "ssl certificate": "path/to/the/crt/file",
    "ssl key": "path/to/the/key/file",
    "ssl ca": "path/to/the/crt/file",
    "timeout": 15
  },

  "files": [
    {
      "paths": [
        "path/to/the/log/files"
      ] ,
      "fields": { "type": "access_log" }
    }, {
      "paths": [ "-" ],
      "fields": { "type": "stdin" }
    }
  ]
}
```

Let's explain just a little bit this configuration:

- Network configuration basically uses an array of target downstream servers that will receive the log files. Each item in the **servers** array is a location to a server, ex: localhost: 5043

  - Then the **ssl*** entries point to the .key and .crt files that we have just generated earlier.

  - Last parameters is the timeout that defines the time that a Logstash-forwarder will wait for a downstream server to wait before assuming the connection is interrupted and then try to connect to another server from the **server's** array.

- The files entry is an array of the data that will be transported; in my example, I'm using:

  - An absolute path to directory that will contain the example log files used for this book. The field value set the event type.

  - A special path contained a "-", which means that the Logstash-forward will forward the standard input content, in other words, whatever you type in the command line prompt. This is a good way to test the transport pipeline between Logstash-forward and Logstash server.

There are more options to configure the files path such as using wildcards that are really practical for rotating log files. More information on the configuration can be found on the Logstash-forwarder Github repository:

https://github.com/elastic/logstash-forwarder

Logstash processing agent also needs a configuration file. It follows a direct structure as mentioned in this chapter. Listing 4-8 shows the one that I'm using for this book.

Listing 4-8. Logstash processing pipeline example, which can be found as: https://github.com/bahaaldinc/scalable-big-data-architecture/blob/master/chapter4/logstash/processor/forwarder_to_kafka.conf

```
input {
  lumberjack {
      port => "5043"
      ssl_certificate => "path/to/crt/file"
      ssl_key => "path/to/key/file"
  }
}
filter {
  grok {
    match => {
      "message" => "%{COMBINEDAPACHELOG}"
    }
  }
}
output {
  stdout { codec => rubydebug }
  kafka {
    topic_id => "clickstream"
  }
}
```

Three parts are present in this configuration file:

- Input: contains the input stream that Logstash listens to, to receive the data. Here we listening on localhost:5043 to our Logstash-forwarder and, as mentioned, we are using the .key and .crt file. We also

- Filter: takes each line contains in the received logs line and transforms them into a structure event. Here we are using grok filter, which is particularly useful when dealing with Apache HTTP server combined log. The grok combined log pattern is a composition of multiple grok pattern. It has the declaration described in Listing 4-9.

Listing 4-9. Grok Apache commong log & combined log pattern

```
COMMONAPACHELOG %{IPORHOST:clientip} %{USER:ident} %{USER:auth} \[%{HTTPDATE:timestamp}\]
"(?:%{WORD:verb} %{NOTSPACE:request}(?: HTTP/%{NUMBER:httpversion})?|%{DATA:rawrequest})"
%{NUMBER:response} (?:%{NUMBER:bytes}|-)

COMBINEDAPACHELOG %{COMMONAPACHELOG} %{QS:referrer} %{QS:agent}
```

The **COMBINEDAPACHELOG** pattern relies on two other patterns: **COMMONAPACHELOG** and **QS** patterns. Indeed, grok comes with a set of predefined patterns that can be reused and composed to create multilevel patterns. You can get more information on the grok filter on this page:

```
https://www.elastic.co/guide/en/logstash/current/
plugins-filters-grok.html
```

And more on the predefined patterns here:

```
https://github.com/elastic/logstash/blob/v1.4.2/patterns/grok-patterns
```

Output: connection to our Apache Kafka server to put the data there. We also used a stdout output with a ruby codec, which is really handy to debug the pipeline and see the structure of the event.

The Message Broker

Now we will configure our Apache Kafka cluster. To do so, we'll create a cluster of two brokers that are exposing a clickstream topic. A topic is basically the channel on which the Logstash processor will publish each events. Then another program will subscribe to the topic and consume the message; in the case of our architecture this will be done either by Logstash or by Apache Spark.

Downloading, installing, and configuring Apache Kafka is pretty straightforward and requires the following steps:

- Start by downloading Kafka on the following page:

  ```
  http://kafka.apache.org/downloads.html
  ```

- Extract and access the config folder under the installation directory to locate the **server.properties** file

- We will duplicate that file and create two separate files for our cluster of two brokers. You should have **server-1.properties** and **server-2.properties**.

- Modify the properties, as Table 4-1 shows, within the two files so there won't be any clashes between the two instances.

Table 4-1.

	server-1.properties	server-2.properties
broker.id	1	2
port	9092	9093
log.dirs	/tmp/kafka-logs-1	/tmp/kafka-logs-2

The three properties respectively represent the broker identifier, the broker listening port, and the broker-logging directory. Both files can be found on the following repo:

https://github.com/bahaaldine/scalable-big-data-architecture/tree/master/chapter4/kafka

We are now ready to connect the dots and run the first part of our architecture to validate it.

Running the Magic

We'll run each block of the shipping part from downstream to upstream; thus, we'll start with Kafka. Kafka relies on Apache Zookeeper especially for managing the cluster of brokers; we'll first launch it from the installation directory with the command shown in Listing 4-10.

Listing 4-10. Zookeper start command

```
bin/zookeeper-server-start.sh config/zookeeper.properties
```

This will output a certain amount of log lines but you should have something that indicates that the Zookeeper server is running, as shown in Listing 4-11.

Listing 4-11. Zookeper starting logs

```
...
[2015-07-11 22:51:39,787] INFO Starting server (org.apache.zookeeper.server.
ZooKeeperServerMain)
...
[2015-07-11 22:51:39,865] INFO binding to port 0.0.0.0/0.0.0.0:2181
(org.apache.zookeeper.server.NIOServerCnxnFactory)
...
```

Now we'll start both brokers' instance, with the commands shown in Listing 4-12.

Listing 4-12. Kafka servers start commands

```
bin/kafka-server-start.sh path/to/server-1.properties &
bin/kafka-server-start.sh path/to/server-2.properties &
```

You should get the lines of logs for the first broker as shown in Listing 4-13.

Listing 4-13. First broker startup log lines

```
[2015-07-11 22:59:36,934] INFO [Kafka Server 1], started (kafka.server.KafkaServer)
[2015-07-11 22:59:37,006] INFO New leader is 1 (kafka.server.ZookeeperLeaderElector$LeaderC
hangeListener)
```

Listing 4-14 shows the log lines for the second broker.

Listing 4-14. Second broker startup log lines

```
[2015-07-11 23:00:46,592] INFO [Kafka Server 2], started (kafka.server.KafkaServer)
```

Listing 4-15 shows how to create a topic in our cluster.

Listing 4-15. Kafka create topic command

```
bin/kafka-topics.sh --create --zookeeper localhost:2181 --replication-factor 3
--partitions 1 --topic clickstream
```

Why is there a replication-factor of 2? Because our cluster will be able to tolerate up to one server failure without losing any data.

To be sure that our shipping part messages are consumable downstream, we'll run a test consumer as shown in Listing 4-16.

Listing 4-16. Kafka test consumer command

```
bin/kafka-console-consumer.sh --zookeeper localhost:2181 --from-beginning –topic clickstream
```

Now let's run the Logstah processing agent, from the Logstash agent installation directory, as shown in Listing 4-17.

Listing 4-17. Logstash processing agent start command

```
/bin/logstash -f /path/to/forwarder_to_kafka.conf
```

which should have the output shown in Listing 4-18.

Listing 4-18. Logstash start logs

```
Logstash startup completed
```

Finally, the Logstash-forwarder is shown in Listing 4-19.

Listing 4-19. Logstash-forwarder start command

```
/Applications/logstash-forwarder/logstash-forwarder -config forwarder.json
```

This should produce the output shown in Listing 4-20.

Listing 4-20. Logstash-forwarder start logs

```
2015/07/12 00:00:29.085015 Connecting to [::1]:5043 (localhost)
2015/07/12 00:00:29.246207 Connected to ::1
```

We are ready to generate an access log file and see the stream through our shipping pipeline, launch the script, and generate one line as shown in Listing 4-21.

Listing 4-21. Log generator start command

```
./generator.py 1
```

Let's check the log on the Logstash processor logs, as shown in Listing 4-22.

Listing 4-22. Logstash processor logs

```
{
    "message" => "10.10.10.15 - - [10/Oct/2013:00:03:47 +0000] \"GET /products/product3
HTTP/1.0\" 401 3627 \"http://www.amazon.com\" \"Mozilla/5.0 (Windows; U; MSIE 9.0;
WIndows NT 9.0; en-US))\" ",
   "@version" => "1",
 "@timestamp" => "2015-07-11T22:19:19.879Z",
       "file" => "/Users/bahaaldine/Dropbox/apress/demo/chapter4/source/
       access_log_20150712-001915.log",
       "host" => "MBPdeBaaaldine2",
     "offset" => "0",
       "type" => "access_log",
   "clientip" => "10.10.10.15",
      "ident" => "-",
       "auth" => "-",
  "timestamp" => "10/Oct/2013:00:03:47 +0000",
       "verb" => "GET",
    "request" => "/products/product3",
"httpversion" => "1.0",
   "response" => "401",
      "bytes" => "3627",
   "referrer" => "\"http://www.amazon.com\"",
      "agent" => "\"Mozilla/5.0 (Windows; U; MSIE 9.0; WIndows NT 9.0; en-US))\""
}
```

As said earlier, the **ruby codec** from the **stdout** output is really practical to see the content of the streamed event. We can see in the previous log that the grok has done its job and parsed the log line and structured the event properly to the expected format.

Now we should also be able to see the same message on the Kafka consumer side but in a JSON format, as shown in Listing 4-23.

Listing 4-23. Kafka consumer log file

```
{
   "message": "10.10.10.15 - - [10\/Oct\/2013:00:03:47 +0000] \"GET \/products\/product3
HTTP\/1.0\" 401 3627 \"http:\/\/www.amazon.com\" \"Mozilla\/5.0 (Windows; U; MSIE 9.0;
WIndows NT 9.0; en-US))\" ",
   "@version": "1",
   "@timestamp": "2015-07-11T22:19:19.879Z",
   "file": "\/Users\/bahaaldine\/Dropbox\/apress\/demo\/chapter4\/source\/access_
log_20150712-001915.log",
   "host": "MBPdeBaaaldine2",
   "offset": "0",
   "type": "access_log",
   "clientip": "10.10.10.15",
   "ident": "-",
   "auth": "-",
   "timestamp": "10\/Oct\/2013:00:03:47 +0000",
   "verb": "GET",
```

```
    "request": "\/products\/product3",
    "httpversion": "1.0",
    "response": "401",
    "bytes": "3627",
    "referrer": "\"http:\/\/www.amazon.com\"",
    "agent": "\"Mozilla\/5.0 (Windows; U; MSIE 9.0; WIndows NT 9.0; en-US))\""
}
```

So far so good; we are sure now that our shipping and processing pipeline works from end to end. Let's now go on the draining side and index our data.

Draining the Logs from Apache Kafka

Draining consists of using a consumer agent, which connects on Kafka and consume the message in the topic. This work will be done again by Logstash but with a different configuration for the agent, as this last one, will index the data in Elasticsearch.

Configuring Elasticsearch

Before draining the data, we need to configure our endpoint, the Elasticsearch cluster.

Installation

Regardless of the method that we'll use to index the data in Elasticsearch, from Logstash or from Spark, we'll need to set up an Elasticsearch cluster. In this book we'll simulate a real-life example and work on a single node cluster, but adding more Elasticsearch node takes a matter of seconds.

Download and extract Elasticsearch from the following link:

```
https://www.elastic.co/downloads/elasticsearch
```

After extracting, you'll need to install a plugin that will help us to have a better a better understanding of what's happening in our Elasticsearch, Marvel. To do so, go in the Elasticsearch extracted folder and run the following command:

```
bin/plugin -i elasticsearch/marvel/latest
```

It's worth mentioning here that the command may vary between the Marvel version used in this book and the latest release. For more information, please refer to the Marvel installation documentation here:

```
https://www.elastic.co/guide/en/marvel/current/installing-marvel.html
```

This will install Marvel in couple of seconds. Then run Elasticsearch:

```
bin/elasticsearch
```

Browse the following URL to the Marvel console and go in the Shard Allocation from the top right menu: http://localhost:9200/_plugin/marvel, as shown in Figure 4-5.

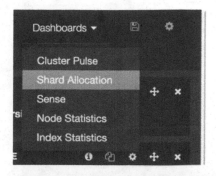

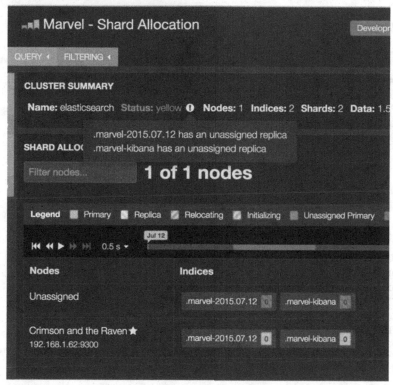

Figure 4-5. Accessing the Shard Allocation dashboard

As you can see, Marvel is mentioning that there are two unassigned replicas in the cluster, which turn the overall status to Yellow. This is completely normal as long as we are running a single instance cluster. Elasticsearch won't assign a replica shard to the same node where the primary shard is hosted; that wouldn't make any sense. Remember that the replica shards provide resiliency to your cluster; in our case we would need to launch another instance of Elasticsearch but as we are in a test environment that would be too much. Let's then change the cluster configuration by using Elasticsearch indices settings API, switch to Sense console from the right menu by clicking on Sense, you will land on the Sense console that help running API queries easily with autocompletion.

What we want to do here is to update the number of replicas for every shard in indeces created in our cluster to 0, and to turn the cluster state to green. Copy and paste the query in Listing 4-24 and click the run button.

Listing 4-24. Indeces API to update the number of replica to 0

```
PUT /_settings
{
    "index" : {
        "number_of_replicas" : 0
    }
}
```

Sense will return an acknowledgement response as shown in Figure 4-6.

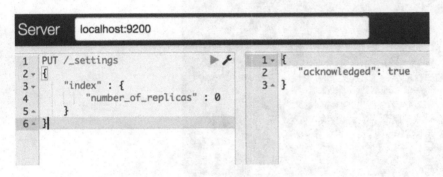

Figure 4-6. *Query result*

As you can see, the request is pretty simple to understand: it uses the settings API and send a PUT request with a body containing an index object to specify that each shard in index should have 0 replicas. If you go back now to the shard allocation view, you should have a green cluster state.

Creating the Index Template

We now will create a mapping for our data, and set the correct parameters for our index, document, and fields. In sense, issue the request described in Listing 4-25.

Listing 4-25. Clickstream indeces template

```
PUT _template/clickstream
{
  "template": "clickstream-*",
  "settings": {
    "number_of_shards": 1,
    "number_of_replicas": 0
  },
  "mappings": {
    "_default_": {
      "dynamic_templates": [
```

```
    {
      "string_fields": {
        "mapping": {
          "index": "not_analyzed",
          "omit_norms": true,
          "type": "string"
        },
        "match_mapping_type": "string",
        "match": "*"
      }
    }
  ],
  "_all": {
    "enabled": true
  },
  "properties": {
    "response": { "type": "integer"},
    "bytes": { "type": "integer" }
  }
}
}
}
}
```

Here we are creating a template called **clickstream** that will be applied to every index created in Elasticsearch which name will math the **clickstream**-* regex, we'll see the reason for this later on in this chapter. It contains the main part:

- The settings: we are defining here the number of shard and replica that we will need in our cluster for the clickstream indices. This is equivalent to what we have done earlier but now packaged in a template and apply to every new clickstream index.

- The mappings: this defined our the data should be analyzed by Elasticsearch. Elasticsearch comes with a very important notion, called analyzer, which basically defined our fields should be interpreted and structure for search purpose. You can for example use a tokenizer and split your field content into multiple token. The analyzers concept is really rich and I would recommend that you read the following documentation for better understanding of what is happening here:

  https://www.elastic.co/guide/en/elasticsearch/reference/1.6/
  analysis-analyzers.html

- I've chosen to not analyze every string field by default so those will basically be searchable as they are. Then I've applied specific mapping to the **response** and **bytes** fields to tell Elasticsearch that these will be integers.

Keep in mind that the mappings should be done before indexation; otherwise, you will have to reindex your entire data.

Indexing the Data Directly with Logstash

Now that Elasticsearch is configured and running, we can launch our Logstash indexer instance and drain Kafka.

The Configuration File

To store the documents created by our clickstream logging architecture, we could create a single index and store everything in it. But as our site traffic grows, the amount of data will grow, so we should consider splitting our index based on time, which gives more sense here.

Think, for example, that the more you are getting data, the more you have old data that might not be useful, which will likely be archive to leave more server resource to the fresher data. The good news is that it's super easy and is done directly at the indexation level, in other words, by Logstash.

Listing 4-26 is the configuration of our Logstash indexer.

Listing 4-26. Logstash indexer configuration and can be found in https://github.com/bahaaldine/ scalable-big-data-architecture/blob/master/chapter4/logstash/indexer/kafka_to_elasticsearch.conf

```
input {
  kafka {
    topic_id => "clickstream"
  }
}
filter {
}
output {
  stdout { codec => rubydebug }
  elasticsearch {
    index => "clickstream-%{+YYYY.MM.dd}"
    manage_template => false
    host => localhost
    protocol => http
  }
}
```

As you can see, the configuration is quite simple compared to the processing agent, and comes with a new output dedicated to index in Elasticsearch with the following settings:

- index: is the index name that we want to index the data into; here the index is pattern and will generate a day based index name

- manage_template: whether or not Logstash should manage the index template; As we have created one earlier this is not needed.

- host and protocol: these are obvious and point to our Elasticsearch cluster

Run and Index the Sata

Let's run our Logstash index and see what happen in Elasticsearch. Issue the command shown in Listing 4-27.

Listing 4-27. Logstash indexer start command

```
bin/logstash -f /path/to/kafka_to_elasticsearch.conf
```

Generate a single event log file with the python script, and, like earlier, you should see the ruby debug codec outputting the event structure. Let's focus on Elasticsearch and go into the Shard Allocation dashboard to see if our clickstream index has been created, as shown in Figure 4-7.

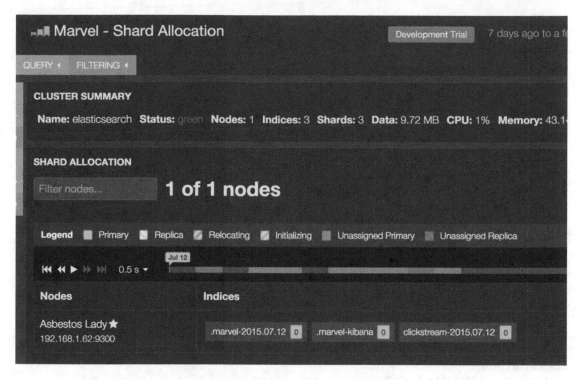

Figure 4-7. *New clickstream index created in Marvel*

We have now a **clickstream-2015.07.12** index that has been created. Obviously, the date might change depending on the time you have launch the Logstash indexer, but anyway you should have one index.

Now let's go into Sense and query our data by issuing the command shown in Listing 4-28.

Listing 4-28. Querying clickstream data in Sense

```
GET /clickstream-2015.07.12/_search
{
  "query": {
    "match_all": {}
  }
}
```

The output you get should be similar to that shown in Listing 4-29.

Listing 4-29. Search output

```
{
   "took": 1,
   "timed_out": false,
   "_shards": {
      "total": 1,
      "successful": 1,
      "failed": 0
   },
   "hits": {
      "total": 1,
      "max_score": 1,
      "hits": [
         {
            "_index": "clickstream-2015.07.12",
            "_type": "access_log",
            "_id": "AU6CJyWNDD4MpZzVyM-9",
            "_score": 1,
            "_source": {
               "message": "10.10.10.13 - - [10/Oct/2013:00:04:55 +0000] \"GET /products/
product3 HTTP/1.0\" 200 3219 \"http://www.google.com\" \"Mozilla/4.0 (compatible; MSIE 7.0;
Windows NT 6.0)\" ",
               "@version": "1",
               "@timestamp": "2015-07-12T12:04:39.882Z",
               "file": "/Users/bahaaldine/Dropbox/apress/demo/chapter4/source/access_
log_20150712-140436.log",
               "host": "MBPdeBaaaldine2",
               "offset": "0",
               "type": "access_log",
               "clientip": "10.10.10.13",
               "ident": "-",
               "auth": "-",
               "timestamp": "10/Oct/2013:00:04:55 +0000",
               "verb": "GET",
               "request": "/products/product3",
               "httpversion": "1.0",
               "response": "200",
               "bytes": "3219",
               "referrer": "\"http://www.google.com\"",
               "agent": "\"Mozilla/4.0 (compatible; MSIE 7.0; Windows NT 6.0)\""
            }
         }
      ]
   }
}
```

We can see that the document has been stored and returned by Elasticsearch and respect the structure we have configured upstream.

Creating a Clickstream Alias

The last thing that I want to configure here in Elasticsearch is an alias. As mentioned, an index will be created based on time, say per day, so we will end up with N indices but we might want to make searches and analyses on all the indices.

To group all of the indices into a single name, we'll create an alias that point on all of our indices. A client application will just need to know the alias name to access the data. Issue the command shown in Listing 4-30 API query in Sense to create our alias.

Listing 4-30. Using the clickstream alias

```
POST /_aliases
{
  "actions" : [
      { "add" : { "index" : "clickstream-*", "alias" : "clickstream" } }
  ]
}
```

Now in Sense, you can issue the command shown in Listing 4-31 and get the exact same result as the previous query.

Listing 4-31. Querying with clickstream alias

```
GET /clickstream/_search
{
  "query": {
    "match_all": {}
  }
}
```

Summary

Our streaming/logging architecture is now ready to ingest data and index them into a distributed durable datastore, for example, Elasticsearch. In the next chapter we'll see how to use Elasticsearch API to query the data in near-real time, and also will see the implementation of the second part of this ingestion architecture: the Spark processing part.

CHAPTER 5

∎∎∎

Querying and Analyzing Patterns

There are different ways to step into data visualization, starting with the analytics strategy that should allow us to identify patterns in real time in our data, and also to leverage the already ingested data by using them in continuous processing. This is what we'll cover in this chapter through the integration of Spark, the analytics in Elasticsearch, and finally visualizing the result of our work in Kibana.

Definining an Analytics Strategy

In this section we'll go through two different approaches for analyzing our data, whether we want to continuously process our data and update the analytics in our document store or make real-time queries as the demand comes. We will see at the end that there is no black-and-white strategy in a scalable architecture; you will likely end up doing both, but you usually start with one of them.

Continuous Processing

What I call continuous processing is the ability to extract KPI and metrics from batch of data as they come. This is a common use case in a streaming architecture that once set up can bring a lot of benefits. The idea here is to rely on a technology, which is:

- Distributed

- Resilient

- Does the processing in near real time

- Is scalable in term of server and processing

Spark is a good candidate to play this role in the architecture; you can basically integrate Spark and Kafka to implement continuous batch processing. Indeed, Spark will consume messages from Kafka topics in batch intervals and apply processing to the data before indexing it into a data store, obviously Elasticsearch:

- The first benefit is to save the datastore from doing the processing

- The second one is to be able to have processing layer such as Spark that can scale. You will be able to add more Spark Kafka consumer that does different kind of processing over time.

The typical processing done is an aggregation of the batch data such as, in our example, number of queries, number of 404 errors, volume of data exchange with users over time, and so on. But, eventually, this processing can reach the point where we do prediction, on data by using the machine learning features of Spark.

Once processed data are delivered by Kafka, streamed and processed by Spark, they should be available for analytics and search. This is where Elasticsearch comes into play and will allow the user to access in real time to data.

Real-Time Querying

Real-time querying removes the burden of relying on a tradional batch processing architecture such as a Hadoop-oriented distribution, which can be heavy to deploy and is not what end users expect.

Today, you typically want the information now, not in five minutes or the next day.

The strategy of real-time queryring is efficient depending on the datastore you are dealing with. It should also provide a set of API that let the users extract essential information from a large set of data at scale.

Elasticsearch does tis: it supports running aggregation queries through the aggregation framework to group document or compute document field data in near real time.

With real-time querying and the out-of-the-box aggregation capabilities of Elasticsearch, we can have an ingestion pipeline that gets rid of any processing between the time that the data have been ingested, to the moment the data are indexed in Elasticsearch. In other words, the pipeline can be reduced to draining the data from Kafka with Logstash and indexing them into Elasticsearch.

The idea here is to see what you should do in Spark and what you should keep for Elasticsearch. We'll see that implementing simple grouping aggregation could be very handy with Elasticsearch.

Process and Index Data Using Spark

To ingest the data with Spark we first need to implement a Spark streaming agent that will consume the data from Kafka and then process and index them in Elasticsearch. Let's then split the ingestion with Spark into two parts: the streamer and the indexer.

■ **Note**   that code used in this chapter can be found in the following repo: `https://github.com/` `bahaaldine/scalable-big-data-architecture/tree/master/chapter5/spark-scala-streamer`.

Preparing the Spark Project

To implement the streaming agent, we'll use Spark Streaming, a reliable, fault-tolerant API that is used to stream data from variant data sources. More information can be found at:

`https://spark.apache.org/streaming/`

So far, we won't structure the data that we receive. The goal here is to set up the streaming itself, ingest the stream from Kafka, and be able to print it out. I'm used to Eclipse (Mars 4.5) to implement my Java (1.8)/ Scala(2.10) project. IntelliJ is also a really good IDE and very popular for Scala applications.

The first thing we need to do is to create a Spark project structure with needed dependencies. To do so, we'll use Maven and a Spark archetype that you can easily find on Github. In my example, I'll use:

```
https://github.com/mbonaci/spark-archetype-scala
```

We this archetype you can simply generate a Scala project that you can then use in Eclipse Execute the command shown in Listing 5-1.

Listing 5-1. Spark Scala Archetype generation

```
mvn archetype:generate \
  -DarchetypeCatalog=https://github.com/mbonaci/spark-archetype-scala/raw/master/
  archetype-catalog.xml \
  -DarchetypeRepository=https://github.com/mbonaci/spark-archetype-scala/raw/master \
  -DgroupId=org.apress.examples \
  -DartifactId=spark-scala-streamer \
  -Dversion=1.0-SNAPSHOT
```

■ **Note**   that the escape character used here to break the line should be adapted to the OS; for example, that would be « ^ » on windows.

Maven will prompt and ask for confirmation prior to generating the project. Go through those steps and finalize the creation. You should then have the directory structure like that shown in Figure 5-1.

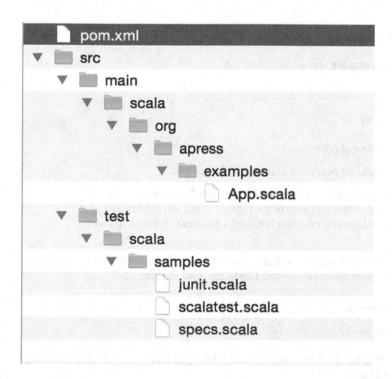

Figure 5-1. *Spark Scala project directory structure*

This directory structure is really important in order for Maven to be able to compile and package your application correctly.

First thing we are going to do is to understand the sample application that Maven has generated for us.

Understanding a Basic Spark Application

The Spark archetype has the benefit of generating a sample application that gives the basic structure of a Spark program. It first starts with the configuration of the Spark context, then handles the data processing, and sends to an output the result of that processing.

In the sample program (App.scala file), the configuration of the SparkContext consists of setting the Application name that will appear in the cluster metadata and the cluster on which we will execute the application, as shown in Listing 5-2.

Listing 5-2. Spark context configuration

```
val conf = new SparkConf()
  .setAppName("The swankiest Spark app ever")
  .setMaster("local[2]")
```

We use the special value of **local[2]** here, which means that we want to run the application on our local machine using a local Spark runtime with two threads. That's really convenient for testing purposes and we'll use this method in this book.

Once the configuration is done, we will create a new Spark configuration using the configuration shown in Listing 5-3.

Listing 5-3. Spark creation

```
val sc = new SparkContext(conf)
```

Then the sample program creates a dataset, shown in Listing 5-4.

Listing 5-4. Spark dataset creation

```
val col = sc.parallelize(0 to 100 by 5)
```

There are actually two steps in this line of code:

- it first creates a collection holding every five numbers from 0 to 100

- Then Spark parallelizes the collection by copying it into a distributed dataset. In other words, and depending on the topology of the cluster, Spark will operate on this dataset in parallel. This dataset is commonly called a Resilient Distributed Data (RDD.

Note that here we don't set the number of partitions that we want to split the dataset into because we let Spark take care of this, but we could have partionned our data like that shown in Listing 5-5.

Listing 5-5. Spark adding partition manually

```
val col = sc.parallelize(0 to 100 by 5, 2)
```

Finally, we transform our data using the Spark sample function, which produces a new RDD containing a new collection of value, as shown in Listing 5-6.

Listing 5-6. Spark creating new RDD

```
val smp = col.sample(true, 4)
```

Here we use a number generator seed of four. If you want to print out the content of the generated sample, use a mirror map function like that shown in Listing 5-7.

Listing 5-7. Spark using mirror map function

```
smp.map(x => x).collect
```

This will print out something like that shown in Listing 5-8.

Listing 5-8. Spark processing output

```
Array(0, 0, 0, 0, 0, 5, 5, 5, 5, 10, 10, 15, 15, 15, 15, 15, 20, 20, 20, 20, 25, 30, 30, 35,
35, 35, 35, 35, 35, 35, 35, 40, 40, 45, 45, 55, 55, 55, 55, 60, 65, 70, 70, 70, 70, 70, 70,
70, 75, 75, 75, 75, 80, 80, 80, 80, 85, 85, 90, 90, 100, 100, 100, 100, 100)
```

If you run the sample application using the Listing 5-9 command:

Listing 5-9. Spark start command

```
mvn scala:run -DmainClass=org.apress.examples.App
```

the last part of the code (Listing 5-10:

Listing 5-10. Spark driver code

```
val colCount = col.count
val smpCount = smp.count
println("orig count = " + colCount)
println("sampled count = " + smpCount)
```

will print out:

Listing 5-11. Spark driver output

```
orig count = 21
sampled count = 78
```

This Spark application is called a Spark driver. Obviously, this one is really easy to understand and uses a few features, but it gives the general idea of a driver structure. What we want to achieve with Spark in our use case is to be able to process a large amount of data, in a distributed way, and digest them before sending them downstream in the architecture.

Implementing the Spark Streamer

Introducing Spark in our existing architecture doesn't require to modify what we have seen in Chapter 4; it's just a matter of adding a new consumer to our Kafka clickstream topic like that shown in Figure 5-2.

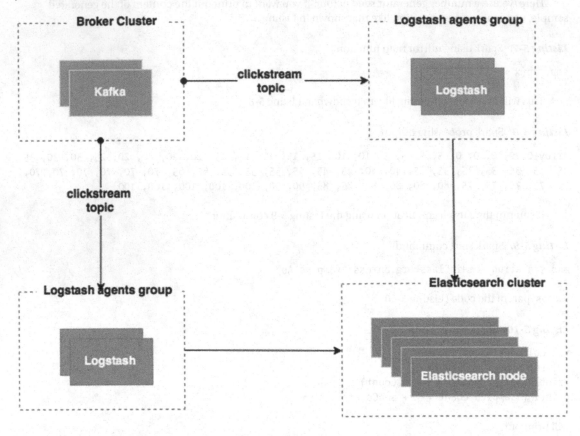

Figure 5-2. *Adding Spark Kafka streaming to our architecture*

Our Spark driver should be able to connect to Kafka and consume the data stream. To do so, we will configure a special Spark context called Streaming context, as shown in Listing 5-12.

Listing 5-12. Configuring a Spark Streaming context

```
val sparkConf = new SparkConf().setAppName("KafkaStreamerToElasticsearch ")
sparkConf.set("es.index.auto.create", "true")
val ssc = new StreamingContext(sparkConf, Seconds(2))
```

As you can see, the context configuration is pretty much the same as that in the previous sample Spark driver, with the following differences:

- We use the **es.index.auto.create** option to make sure that the index will be created automatically in Elasticsearch when the data is indexed.

- We set the StreamingContext at a 2-second batch interval, which means that data will be streamed every 2 seconds from the Kafka topic.

We need now to configure from what brokers and what topics we want to receive the data, and then create the stream line between the two worlds, as shown in Listing 5-13.

Listing 5-13. Connecting Spark and Kafka

```
val Array(brokers, topics) = args
val topicsSet = topics.split(",").toSet
val kafkaParams = Map[String, String]("metadata.broker.list" -> brokers)
val messages = KafkaUtils.createDirectStream[String, String, StringDecoder, StringDecoder]
(ssc, kafkaParams, topicsSet)
```

The **args** variable contains the arguments passed to the driver when running it. The list of brokers and list of topics are both delimited by a comma. Once parsed, the KafkaUtils class, provided as part of the Spark streaming framework (org.apache.spark.streaming package), is used to create a direct stream between Kafka and Spark. In our example all messages are received in the **messages** variable.

At this point we should be able to output each streamed clickstream log line by using the map function in Listing 5-14.

Listing 5-14. Map and print lines

```
val lines = messages.map(_._2)
lines.print();
```

There are a couple of things to examine in the previous lines:

- the messages variables is an RDD that holds the streamed data
- the _ syntax represents an argument in an anonymous function, as in Listing 5-15.

Listing 5-15. Spark map function simplified syntax

```
map(_._2)
```

This is equivalent to the line shown in Listing 5-16, where **_2** is the second element of the x tuple.

Listing 5-16. Spark map function full syntax

```
map(x => x._2)
```

To run the driver, we'll first build it using maven. Run the following command to compile and package the Scala code into a JAR:

```
mvn scala:compile package
```

Note that in the case of that driver, I'm using Scala 2.10.5, which requires Java 7 in order to work:

```
export JAVA_HOME=/Library/Java/JavaVirtualMachines/jdk1.7.0_21.jdk/Contents/Home/
```

Once compiled and packaged, you should end up with a two JARs in the project target directory:

- spark-scala-streamer-1.0.0.jar: contains only the compiled project classes

- spark-scala-streamer-1.0.0-jar-with-dependencies.jar contains compiled classes and project dependencies

We'll use a slightly different method than in the previous part to run the driver: the **spark-submit** executable from the standard spark distribution, which can be downloaded from here:

```
http://spark.apache.org/downloads.html
```

At the time of the writing I was using the 1.4.0 version. Once downloaded I recommend that you simply add Spark binaries to your PATH environment variable, as shown in Listing 5-17.

Listing 5-17. Adding Spark binaries to PATH environment variable

```
export SPARK_HOME=/Applications/spark-1.4.0/bin
export PATH=$SPARK_HOME:$PATH
```

This will ease the access to all Spark executables including spark-submit that you can use like that in Listing 5-18 to execute our previous program.

Listing 5-18. Running the Spark Driver using spark-submit

```
spark-submit --class org.apress.examples.chapter4.KafkaStreamer \
  --master local[1] \
  target/spark-scala-streamer-1.0.0-jar-with-dependencies.jar \
localhost:9092,localhost:9093 clickstream
```

As you can see, we pass the KafkaStreamer to spark-submit, which is part of the JAR we just created. We are using the JAR with dependencies because otherwise Spark will not find the needed class and you will likely end up with a "class not found" exception. At the end of the command, you have probably recognized the argument containing the broker list that we want to connect to and the topic that we want to stream data from. Running this command just bootstraps Spark and listen for incoming message, thus generate log lines using the generator we used in the previous and be sure that the streaming pipeline is running in order to push message in Kafka. You should get some lines outputted in the console like those in Listing 5-19.

Listing 5-19. Example of data streamed by Spark

```
-------------------------------------------
Time: 1437921302000 ms
-------------------------------------------
{"message":"10.10.10.11 - - [10/Oct/2013:00:04:02 +0000] \"GET /page3 HTTP/1.0\" 200
4401 \"http://www.amazon.com\" \"Mozilla/5.0 (Windows; U; Windows NT 6.1; rv:2.2)
Gecko/20110201\" ","@version":"1","@timestamp":"2015-07-26T14:35:00.869Z","file":"/tmp/
source/access_log_20150726-163457.log","host":"ea8fceb4f5b0","offset":"0","type":"acce
ss_log","clientip":"10.10.10.11","ident":"-","auth":"-","timestamp":"10/Oct/2013:00:04:02
+0000","verb":"GET","request":"/page3","httpversion":"1.0","response":"200","bytes":"4401",
"referrer":"\"http://www.amazon.com\"","agent":"\"Mozilla/5.0 (Windows; U; Windows NT 6.1;
rv:2.2) Gecko/20110201\""}
```

{"message":"10.10.10.15 - - [10/Oct/2013:00:05:44 +0000] \"POST /page6 HTTP/1.0\" 200
2367 \"http://www.bing.com\" \"Mozilla/5.0 (Linux; U; Android 2.3.5; en-us; HTC Vision
Build/GRI40) AppleWebKit/533.1 (KHTML, like Gecko) Version/4.0 Mobile Safari/533.1\"
","@version":"1","@timestamp":"2015-07-26T14:35:00.869Z","file":"/tmp/source/access_
log_20150726-163457.log","host":"ea8fceb4f5b0","offset":"167","type":"access_log","clienti
p":"10.10.10.15","ident":"-","auth":"-","timestamp":"10/Oct/2013:00:05:44 +0000","verb":"P
OST","request":"/page6","httpversion":"1.0","response":"200","bytes":"2367","referrer":"\"
http://www.bing.com\ "","agent":"\"Mozilla/5.0 (Linux; U; Android 2.3.5; en-us; HTC Vision
Build/GRI40) AppleWebKit/533.1 (KHTML, like Gecko) Version/4.0 Mobile Safari/533.1\""}

Now that we have the streaming part of our code running, we can focus on the indexing part.

Implementing a Spark Indexer

The indexing part is the last part of our driver but, surprisingly, we are not pushing its configuration to the last step.

What we will do here is just make sure that each clickstream event that we consume can be indexed in ElasticSearch. The clickstream events are actually JSON received in string stream; we then need to parse the string, marshal to JSON RDD, and then create a RDD of Clickstream object.

Let's first defined our Clickstream object as shown in Listing 5-20.

Listing 5-20. Clickstream object

```
case class Clickstream (
    message:String,
    version:String,
    file:String,
    host:String,
    offset:String,
    eventType:String,
    clientip:String,
    ident:String,
    auth:String,
    timestamp:String,
    verb:String,
    request:String,
    httpVersion:String,
    response:String,
    bytes:Integer,
    referrer:String,
    agent:String.
)
```

Except for the bytes, we assume that every field are Strings. Let's now parse our lines of JSON String as shown in Listing 5-21.

Listing 5-21. Spark parse line

```
val parsedEvents =
lines.map(JSON.parseFull(_))
    .map(_.get.asInstanceOf[scala.collection.immutable.Map[String,Any]])
```

Lines are processed in two steps:

- each lines is parsed using JSON.parseFull
- for each JSON, RDD are transformed into a map of key and value.

Finally, clickstream events RDD are pulled out from the parsedEvents variable as shown in Listing 5-22.

Listing 5-22. Spark Clickstream object parsing

```
val events = parsedEvents.map(data=>Clickstream(
    data("message").toString
    ,data("@version").toString
    ,data("file").toString
    ,data("host").toString
    ,data("offset").toString
    ,data("type").toString
    ,data("clientip").toString
    ,data("ident").toString
    ,data("auth").toString
    ,data("timestamp").toString
    ,data("verb").toString
    ,data("request").toString
    ,data("httpversion").toString
    ,data("response").toString
    ,Integer.parseInt(data("bytes").toString)
    ,data("referrer").toString
    ,data("agent").toString
))
```

We simply make a mapping between each parsed event Map and new Clickstream event object. Then at the end we index the data using the Elasticsearch Hadoop connector as shown in Listing 5-23.

Listing 5-23. Clickstream events indexed into Elasticsearch

```
events.foreachRDD{ rdd =>
if (rdd.toLocalIterator.nonEmpty) {
        EsSpark.saveToEs(rdd, "spark/clickstream")
    }
}
```

I'm creating a different index than the one created by Logstash in the previous chapter. The data are in the **spark** index under the **clickstream** type.

You can find more information on the Elasticsearch Hadoop connector at this link:

```
https://www.elastic.co/guide/en/elasticsearch/hadoop/current/index.html
```

At this point you should be able to index in Elasticsearch after generating data, like in Figure 5-3.

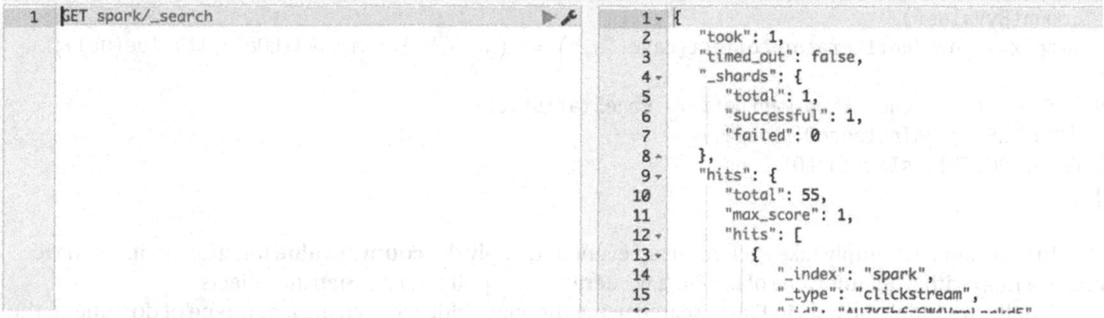

```
 1  GET spark/_search                       ▶ ⚙     1 ▾ {
                                                    2      "took": 1,
                                                    3      "timed_out": false,
                                                    4 ▾    "_shards": {
                                                    5        "total": 1,
                                                    6        "successful": 1,
                                                    7        "failed": 0
                                                    8 ▴    },
                                                    9 ▾    "hits": {
                                                   10        "total": 55,
                                                   11        "max_score": 1,
                                                   12 ▾      "hits": [
                                                   13 ▾        {
                                                   14            "_index": "spark",
                                                   15            "_type": "clickstream",
                                                                 "_id": "AU7VEhf-CWAVml-okdE"
```

Figure 5-3. *Clickstream events in the spark index*

As you can see, I'm using Elasticsearch search API to fetch data from the spark index; as a result, I'm getting 55 clickstream events from the index.

Implementing a Spark Data Processing

That may look weird to end with part, but I prefer that the streaming to pipe be ready to work before diving on the core usage of Spark, the data processing.

We will now add a little bit more processing logic to our previous Spark code and introduce a new object: the PageStatistic object. This object will hold all the statistics that we want to have around user behavior on our website. We'll limit the example to a basic one; in the next chapter, we'll will go deeper.

The processing part will be then be divided into the following parts:

- creation on a simple PageStatistic object that holds the statistics

- computation of statistic

- indexation part

Let's first begin with the PageStatistic object, which so far will only hold the count of GET and POST request, as shown in Listing 5-24.

Listing 5-24. PageStatistic object definition

```
case class PageStatistic (
  getRequest:Long,
  postRequest:Long
)
```

Then we'll use a special from Spark scala library call **countByValue** to compute the number of GET and POST in the batch of logs we receive from Kafka as seen in Listing 5-25.

Listing 5-25. Compute count of GET and POST using countByValue

```
val requestMap = events
  .map(event => event.verb)
  .countByValue()
  .map(x=>x.productIterator.collect{case (a,b) => (a -> b)}.toMap.withDefaultValue(0L));

val requests = requestMap.map(data => PageStatistic(
  data("GET").asInstanceOf[Long],
  data("POST").asInstanceOf[Long]
));
```

In this code, we simply take each received event and apply the **countByValue** function on it. Then we create a map with a default value of 0, which we iterate on to pull out PageStatistic objects.

Finally, we index the data in Elasticsearch using the spark index and create a new type of document, the stats, as seen in Listing 5-26.

Listing 5-26. Indexing stats documents in spark Elasticsearch index

```
requests.foreachRDD{ rdd =>
  if (rdd.toLocalIterator.nonEmpty) {
    EsSpark.saveToEs(rdd, "spark/stats")
  }
}
```

If we check in Elasticsearch using Listing 5-27:

Listing 5-27. Getting spark stats document from Elasticsearch

```
GET spark/stats/_search
```

We get:

```
{
   "took": 1,
   "timed_out": false,
   "_shards": {
      "total": 1,
      "successful": 1,
      "failed": 0
   },
   "hits": {
      "total": 1,
      "max_score": 1,
      "hits": [
```

```
{
    "_index": "spark",
    "_type": "stats",
    "_id": "AU7ReZIVvtspbVGliqwM",
    "_score": 1,
    "_source": {
        "getRequest": 34,
        "postRequest": 12
    }
}
]
}
}
```

We can see that our document holds the two computed fields of the PageStatistic object. But this level of trivial computation is something that Elasticsearch can easily do; we'll see how in the next section. Remember that we'll try to emphasize this Spark processing pipeline in the following chapter, to show the full power of this stack.

Data Analytics with Elasticsearch

Introduction to the aggregation framework

Elasticsearch comes with a powerful set of API that let the users get the best of their data. The aggregation framework groups and enables real-time analytics queries on small to large set of data.

What we have done with Spark in terms of simple analytics can be done as well in Elasticsearch using the aggregation framework at scale. This framework is devided into two types of aggregations:

- Bucket aggregation that aims to group a set of document based on key common to documents and criterion. Document that meets the condition falls in the bucket.

- Metric aggregation that aims to calculate metrics such as average, maximum, minimum, or even date histograms, over a set of documents.

Indeed, the power of aggregation resides on the fact that they can be nested

Bucket Aggregation

Bucket aggregation is used to create groups of documents. As opposed to metric aggregation, this can be nested so that you can get real-time multilevel aggregation. Whereas in Spark creating the PageStatistic structure needed a bunch of coding and logic that would differ from a developer to the other, which could affect the performance of the processing dramatically, in Elasticsearch there is less work to reach the same target.

The bucketing aggregation is really easy to implement. Let's take an example of verb distribution; if we want to group document per verb, all we have to do is follow the code shown in Listing 5-28.

Listing 5-28. Term aggregation on clickstream http verb

```
curl -XGET "http://192.168.59.103:9200/clickstream-2015.07.27/_search" -d'
{
  "size": 0,
  "aggs": {
    "by_verb": {
      "terms": {
        "field": "verb"
      }
    }
  }
}'
```

Which then returns the code shown in Listing 5-29.

Listing 5-29. Term aggregation result

```
{
    "took": 11,
    "timed_out": false,
    "_shards": {
        "total": 1,
        "successful": 1,
        "failed": 0
    },
    "hits": {
        "total": 29280,
        "max_score": 0,
        "hits": []
    },
    "aggregations": {
        "by_verb": {
            "doc_count_error_upper_bound": 0,
            "sum_other_doc_count": 0,
            "buckets": [
                {
                    "key": "get",
                    "doc_count": 24331
                },
                {
                    "key": "post",
                    "doc_count": 4949
                }
            ]
        }
    }
}
```

As you can see, I'm running the aggregation in couple of milliseconds, so near real time (11 ms) and getting the distribution of get and post HTTP verbs. But what if we want to get an equivalent result that we had in Spark, a time-based distribution of HTTP verbs? Let's use the date histogram aggregation and have a nested term aggregation into it, as shown in Listing 5-30.

Listing 5-30. Date histogram with nested term aggregation

```
curl -XGET "http://192.168.59.103:9200/clickstream-2015.07.27/_search" -d'
{
  "size": 0,
  "aggs": {
    "over_time": {
      "date_histogram": {
        "field": "@timestamp",
        "interval": "minute"
      },
      "aggs": {
        "by_verb": {
          "terms": {
            "field": "verb"
          }
        }
      }
    }
  }
}'
```

The date histogram aggregation is indeed really easy to use; it specifies:

- The field holding the timestamp that the histogram will use

- The interval for bucketing. Here I've used minute because I'm indexing more than 1 million documents on the laptop while I'm writing this book. This obviously takes time so I'm sure to have document distributed over minutes. I could have used day, month, or year, or specified a date expression **2h** for "every 2 hours".

You see the nested term aggregation in my query; let's just switch the interval to 20m for getting a more condensed result, shown in Listing 5-31.

Listing 5-31. Date histogram and nested term aggregation result

```
{
  "took": 73,
  "timed_out": false,
  "_shards": {
    "total": 1,
    "successful": 1,
    "failed": 0
  },
  "hits": {
    "total": 228387,
    "max_score": 0,
    "hits": []
  },
```

```
"aggregations": {
    "over_time": {
        "buckets": [
            {
                "key_as_string": "2015-07-27T11:20:00.000Z",
                "key": 1437996000000,
                "doc_count": 105442,
                "by_verb": {
                    "doc_count_error_upper_bound": 0,
                    "sum_other_doc_count": 0,
                    "buckets": [
                        {
                            "key": "get",
                            "doc_count": 87593
                        },
                        {
                            "key": "post",
                            "doc_count": 17849
                        }
                    ]
                }
            },
            {
                "key_as_string": "2015-07-27T11:40:00.000Z",
                "key": 1437997200000,
                "doc_count": 122945,
                "by_verb": {
                    "doc_count_error_upper_bound": 0,
                    "sum_other_doc_count": 0,
                    "buckets": [
                        {
                            "key": "get",
                            "doc_count": 102454
                        },
                        {
                            "key": "post",
                            "doc_count": 20491
                        }
                    ]
                }
            }
        ]
    }
}
```

As you can see, we get two buckets of 20 minutes data and the nested distribution of HTTP verbs. This kind of aggregation is ideal when you want to draw them as timelines or bar charts that render the data over time with a fine control of the period of time you want to analyze the data.

Let's now see what happen if we use a metric aggregation.

Metric Aggregation

Metric aggregations are the last level of set of aggregation and are use to compute metrics over set of documents. Some of the computations are considered as single value metric aggregation because they basically compute a specific operation across document such as average, min, max or sum, some are multi-value aggregation because they give more than one statistic metrics for a specific field across documents such as the stats aggregation.

As an example of multivalue aggregation metric, let's make a stats aggregation on the bytes field to get more visibility on the throughput of our website page, with Listing 5-32.

Listing 5-32. Stats aggregation on the bytes field

```
GET clickstream-2015.07.27/_search
{
  "size": 0,
  "aggs": {
    "by_stats": {
      "stats": {
        "field": "bytes"
      }
    }
  }
}
```

The stats aggregation only takes a numeric field as input—here, bytes—and returns Listing 5-33.

Listing 5-33. Stats aggregation query output

```
{
    "took": 1,
    "timed_out": false,
    "_shards": {
       "total": 1,
       "successful": 1,
       "failed": 0
    },
    "hits": {
       "total": 2843774,
       "max_score": 0,
       "hits": []
    },
    "aggregations": {
       "by_stats": {
          "count": 2843774,
          "min": 2086,
          "max": 4884,
          "avg": 3556.769230769231,
          "sum": 3932112564
       }
    }
}
```

The results give the computation of differents metrics for the bytes field:

- The count represents the number of documents the query has been on, here more than 2.5 million

- The min & max are respectively the minimum and maximum of bytes

- The avg is the average bytes value for all documents

- The sum is the sum of all bytes field across documents

This is how you can with a small query extract statistic from your document, but let's try to go further and combine our bucket and metric aggregation. What we will try to do is to add another level of aggregation and check what is the throughput on each of HTTP verb type, shown in Listing 5-34.

Listing 5-34. Three-level aggregation

```
GET clickstream-2015.07.27/_search
{
  "size": 0,
  "aggs": {
    "over_time": {
      "date_histogram": {
        "field": "@timestamp",
        "interval": "20m"
      },
      "aggs": {
        "by_verb": {
          "terms": {
            "field": "verb"
          },
          "aggs": {
            "by_bytes": {
              "stats": {
                "field": "bytes"
              }
            }
          }
        }
      }
    }
  }
}
```

We will get the bytes statistics for each HTTP verb every 20 minutes, as shown in Listing 5-35.

Listing 5-35. Three aggregation level results

```
{
    "took": 2,
    "timed_out": false,
    "_shards": {
        "total": 1,
        "successful": 1,
        "failed": 0
    },
    "hits": {
        "total": 39,
        "max_score": 0,
        "hits": []
    },
    "aggregations": {
        "over_time": {
            "buckets": [
                {
                    "key_as_string": "2015-07-27T14:00:00.000Z",
                    "key": 1438005600000,
                    "doc_count": 19,
                    "by_verb": {
                        "doc_count_error_upper_bound": 0,
                        "sum_other_doc_count": 0,
                        "buckets": [
                            {
                                "key": "get",
                                "doc_count": 14,
                                "by_bytes": {
                                    "count": 14,
                                    "min": 2086,
                                    "max": 4728,
                                    "avg": 3232.4285714285716,
                                    "sum": 45254
                                }
                            },
                            {
                                "key": "post",
                                "doc_count": 5,
                                "by_bytes": {
                                    "count": 5,
                                    "min": 2550,
                                    "max": 4741,
                                    "avg": 3382.2,
                                    "sum": 16911
                                }
                            }
                        ]
                    }
                },
                {
```

```
            "key_as_string": "2015-07-27T14:20:00.000Z",
            "key": 1438006800000,
            "doc_count": 20,
            "by_verb": {
                "doc_count_error_upper_bound": 0,
                "sum_other_doc_count": 0,
                "buckets": [
                    {
                        "key": "get",
                        "doc_count": 16,
                        "by_bytes": {
                            "count": 16,
                            "min": 2487,
                            "max": 4884,
                            "avg": 3804.25,
                            "sum": 60868
                        }
                    },
                    {
                        "key": "post",
                        "doc_count": 4,
                        "by_bytes": {
                            "count": 4,
                            "min": 3182,
                            "max": 4616,
                            "avg": 3920.25,
                            "sum": 15681
                        }
                    }
                ]
            }
        }
    }
    ]
    }
    }
}
```

The point here is not to redo what Spark does, for this type of operation Elasticsearch is handier; the goal is to use those features to leverage what Spark can do.

Visualize Data in Kibana

As an example of a resulting report, we will create a Kibana 4 dashboard that shows how Elasticsearch aggregation API is used.

The Kibana UI is accessible on your local installation via the following URL:

```
http://localhost:5601
```

In Kibana discovery tab, we can see that we are receiving data, as shown in Figure 5-4.

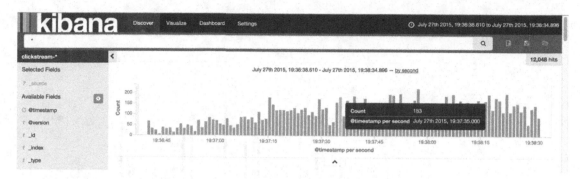

Figure 5-4. Discovering clickstream data

Let's try to make a bar chart visualization that shows the distribution of HTTP response code over time. This can lead to anomaly detection such as intrusion attempt on our website and so on.

The visualization tab lets you create by configuration this kind of charts, as shown in Figure 5-5.

Figure 5-5. Creating aggregation in Kibana

What we expect is for Kibana to draw a report that shows the number of 200, 404, and so on over time, as shown in Figure 5-6.

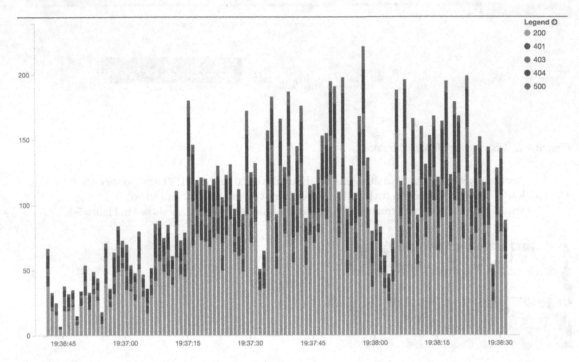

Figure 5-6. *HTTP response code over time*

Visualizing our data brings us to another dimension and easily shows the power of the aggregation framework. Kibana doesn't compute anything on its side; it just renders the data that are returned from Elasticsearch. Listing 5-36 shows what Kibana is asking under the hood.

Listing 5-36. Kibana request to Elasticsearch

```
"aggs": {
    "2": {
      "date_histogram": {
        "field": "@timestamp",
        "interval": "1s",
        "pre_zone": "+01:00",
        "pre_zone_adjust_large_interval": true,
        "min_doc_count": 1,
        "extended_bounds": {
          "min": 1438022198610,
          "max": 1438022314896
        }
      },
      "aggs": {
        "3": {
          "terms": {
```

```
          "field": "response",
          "size": 5,
          "order": {
            "_count": "desc"
          }
        }
      }
    }
  }
}
```

Kibana is relying on Elasticsearch scaling ability and those kinds of visualizations can be done on small to very large sets of data. In the end, what you want to get is a complete view of your data, such as the Kibana dashboard shown in Figure 5-7.

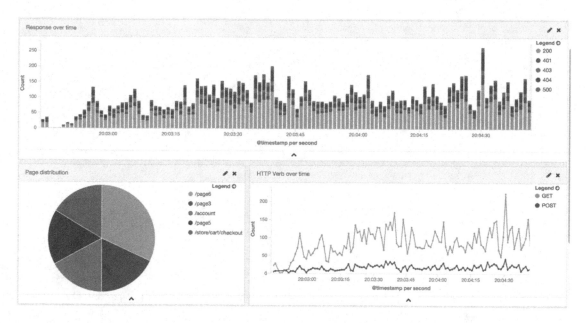

Figure 5-7. *Kibana dashboard*

Summary

In the next chapter, we'll add more complexity to our processing and bring more prediction to it. In addition, we'll continue to illustrate the results of processing with Kibana dashboard.

CHAPTER 6

■ ■ ■

Learning From Your Data?

One of the powerful features of Spark is its machine learning capabilities. Using the Spark Mllib it's very easy to enrich existing data processing pipelines with learning algorithms. This chapter is an introduction to the main concepts behind machine learning and an introduction to augmenting our architecture with machine learning features.

Introduction to Machine Learning

Machine learning is the concept of creating computer programs that can learn from themselves when exposed to new data.

Machine learning is widely used today, and present in almost all industries, to improve user navigation depending on previous behavior, to prescribe relevant drugs or therapy depending on patient background, or to help the Mars robot to get better land navigation as it moves and crosses roadblocks.

As you google something, listen to your Spotify playlist, browse Amazon product catalog, or rent an Airbnb apartment, machine learning is used.

The purpose of this chapter is not to give a course on machine learning, but talk about it in the context of our lambda architecture. However, we will have a quick introduction and talk about supervised and unsupervised learning, which are the most common practices in machine learning, as well as a walkthrough in the algorithm we'll use later: K-means.

Supervised Learning

When dealing with machine learning, you must deal with questions such as which model to use and what strategy to employ to train data? The difference between supervised and unsupervised learning resides in the causal nature of the model.

In supervised learning the inputs, which are a set of oberservations, affect the outputs, which are another set of observations, based on the model described in Figure 6-1.

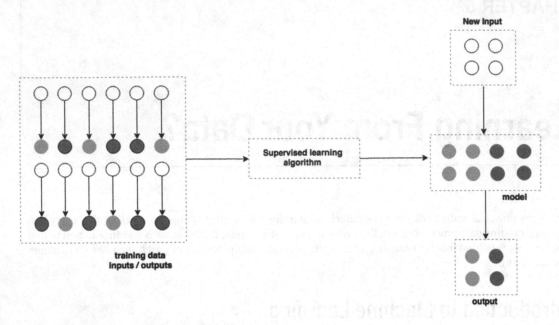

Figure 6-1. *Supervised learning algorithm*

A model is created based on the training data, and is enriched as new data arrives. This process is continuously running.

A real-life example could be training data in order to approve a loan for an individual. By having information about a former enquirer, financial wealth, health, career, and so on, the learning process can deduce whether the loan should be approved or not.

A concrete usage of supervised learning is illustrated in Figure 6-2, showing the average monthly rent cost of an apartment in the center of Paris over one year. A data sample has helped to apply a linear regression on data to reveal a forecast. The red line in Figure 6-2 shows, for example, that the average price will reach €4000 in 2018.

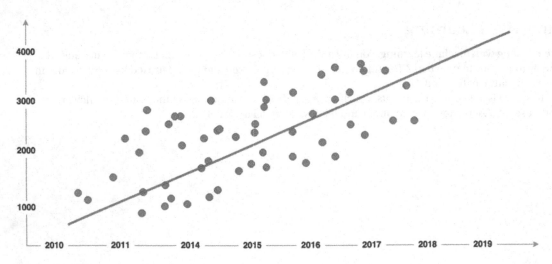

Figure 6-2. *Linear regression based supervised learning*

Such machine learning works really well when dealing with historical data to predict short- to mid-term futures, but these are based on a model that are not obvious in some use cases.

Unsupervised Learning

As opposed to supervised learning, the model in unsupervised learning is created based on the input data and is improved as the data comes in. Some unsupervised algorithms detect similarity in data and tries to create clusters of similar data, such as that seen in Figure 6-3.

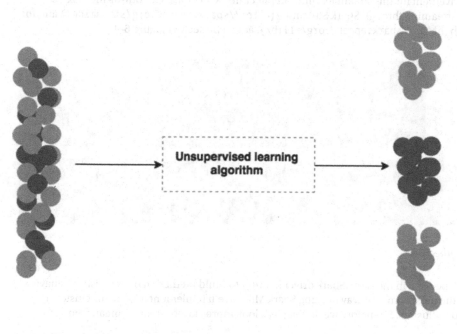

Figure 6-3. *Unsupervised algorithm*

This fits a lot better in our streaming use case where data are not known at the beginning and may have patterns that change over time.

This kind of algorithm is called a clustering algorithm. Some well-known algorithms include:

- K-means

- K-medians

- Expectation maximizations

- Hierachical clustering

In our case, we'll focused on the K-means.

Machine Learning with Spark

It's obvious that machine learning starts with static data. A dataset is created; we build a model on this dataset, then we apply the model to the following created data to come up with a prediction. In our case, when we launch our Spark streaming driver, we don't have any data on which we can base any prediction; we have to build and train the model with the batch of streamed data.

The behavior pattern within our data can change from time to time; that's why streaming architectures fit better in such use cases than bulk batch machine learning. With streaming, we learn from the data and train our model to adapt our prediction over time.

Spark is really convenient for this stream learning architecture as it comes with out-of-the-box components both for streaming through Spark Streaming (`http://spark.apache.org/streaming/`) and for learning through Mllib (`http://spark.apache.org/mllib/`), as can be seen in Figure 6-4.

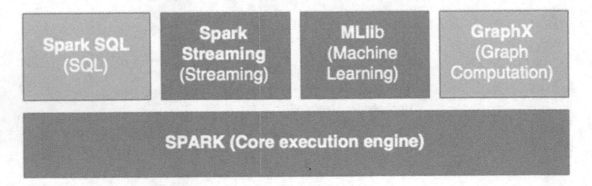

Figure 6-4. *Spark architecture*

Both libraries will be used in the same Spark driver in order to build prediction on our visitor behavior. We'll focus on one of the most common way of using Spark Mllib: we'll implement a K-means clustering to partition our data into K-clusters. But before we do that, let's look more closely at this K-mean algorithm.

Adding Machine Learning to Our Architecture

The K-means algorithm is part of the unsupervised machine learning algorithm. It's one of the most used when it comes to data clustering, and also one of the simplest.

K-means allows us to classify a given data set in a K cluster; each cluster has a center called the centroid. For a given data, we define K centroids, possibly as far as possible from each other, then associate the data point to the nearest centroid to make an early groupage of the data.

Each centroid needs to be recalculated in conjunction with the previous data points that have joined the cluster, which will give K new centroids or cluster barycenters. After recalculating each centroid, all data points are again associated to the nearest centroid. And the recalculation is done again until the centroids don't move any more.

Here are the main steps illustrated by following figures:

1. We first need a dataset:

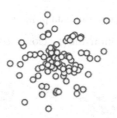

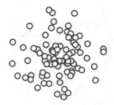

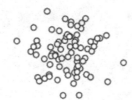

Figure 6-5. *Dataset to be clustered*

2. We then place K centroids in the two-dimensional space to represent our cluster center:

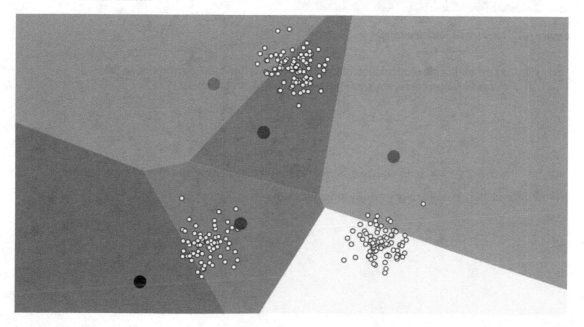

Figure 6-6. *Centroids added to represent clusters*

This splits the space into colored regions, with each color associated to the centroid color. Each data points is associated with the closest centroid, which gives us the early groupage.

3. Then we start recalculating the centroid position relative to the data points:

    As you can see in Figure 6-7, the centroids changed position and got closer to the data points clumps.

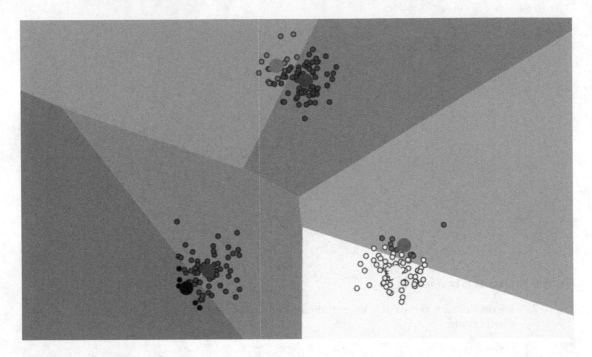

Figure 6-7. *First recalculation phase*

4. At the end of the recalculation phase, the centroids don't move any more, as the following objective function has been minimize to its maximum:

$$J = \sum_{j=1}^{k} \sum_{i=1}^{n} \| x_i^j - c_j \|^2$$

where $\left\| x_j^i - c_j \right\|^2$ is the square distance between a data point x_i^j and the closest centroid c_j.

At the end, the clustering looks slightly different, as shown in Figure 6-8.

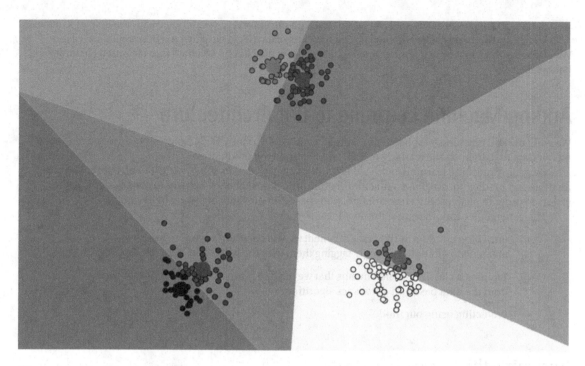

Figure 6-8. *Final clustering state*

For visualizing data I'm using this awesome and very pratical website: http://www.naftaliharris.com/blog/visualizing-k-means-clustering/

Figure 6-8 may make more sense if we add a bit more reality to it:

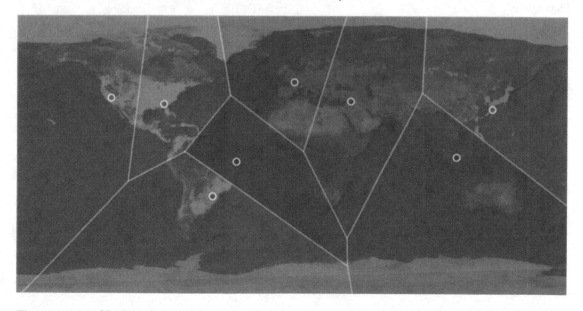

Figure 6-9. *World split with K-means*

Look how you could easily split the world based on the criteria of your choice with the help of K-means!

So I guess this makes more sense now when we correlate this to our current clickstream use case. We want to build a machine learning pipeline that clusters the clickstream data into classified customer behavior.

Adding Machine Learning to Our Architecture

We will now add machine learning to our lambda architecture. This will be progressive as we need to work on our data first, then on the learning model, and finally on the prediction.

By adding machine learning to the architecture, what we want to achieve is user behavior real-time analysis and prediction. We'll use a machine learning pipeline that we'll help up to classify our website vistors in order to push product recommendations based on previous browsing behavior.

The general idea for processing this includes:

- Enriching our data as they are streamed: we will basically add more business meaning to our data by simply tagging them with the help of an extra data source.

- Training a classifier, which means that we need to setup a model in order to learn from the data based on a statistics algorithm.

- Predicting using our model.

Enriching the Clickstream Data

Before diving into pure machine learning we need to add more business sense to our use case. We'll use Spark to achieve this task and leverage the highly resilient and distributed nature of the technology. The strategy is to load a map of data that represents a mapping between the browser URI and the product category, which basically will give us the category of the product browsed by the visitor, as shown in Figure 6-10.

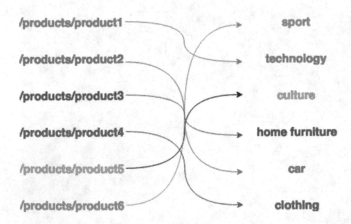

Figure 6-10. *Product mapping*

This is indeed a very short mapping example, which can be loaded and held in memory in Spark without expecting any overhead. In a real-life use case, the mapping would probably have a significant size that would not be safe to hold in memory. In this case, I would use a persistent store such as HDFS, or even, in Elasticsearch. The reason to store it in ES would probably be to ease the search of the appropriate mapping. In this example, we will a create simple CSV file.

The process of enriching the data is shown in Figure 6-11.

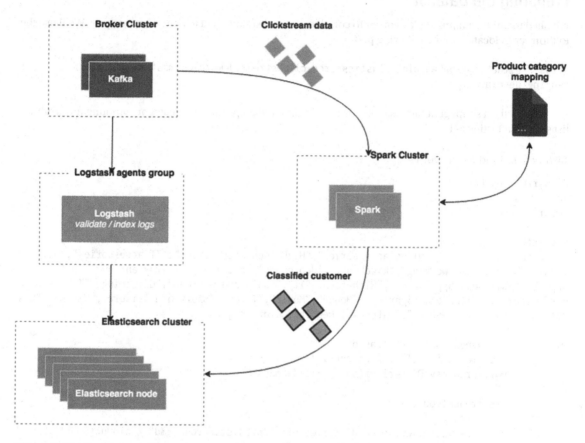

Figure 6-11. *Enrichment diagram*

The data are like those discussed in the previous chapter fetched in Spark, but the difference here is that we'll:

- create a customer object
- compare those data with ours in memory mapping
- enrich our customer objects
- index them in Elasticsearch

To do so, we'll create another Scala project called spark-enrich-and-ml, which be found in my Github account:

```
https://github.com/bahaaldine/spark-enrich-and-ml
```

Preparing the Dataset

We need to build a mapping file in order to correlate the clickstream data to a product category. We'll use the Python script located on the following path:

```
https://github.com/bahaaldine/elasticsearch-clickstream-demo/blob/master/generator/
mapping_generator.py
```

This Python script generates 200,000 URIs and randomly maps them to a category amoung the arbitrary list shown in Listing 6-1.

Listing 6-1. Product categories

```python
#!/usr/bin/python
import csv
import random

productCount = 200000
productCategories = ["Devices accessories","High-Tech accessories","IT accessories","pet"
,"car","baby","home working","kitchen","video games","cycling","blu-ray and dvd","office
supply","household applicance","High-tech","IT","Audio and musique","Gardening","Toys and ga
mes","Books","Software","Lighting","House","Music","Tires","Sport and leisure","Videos","Cos
metics","Jewellery","Shoes","Watches","Heatlh","Clothes"]

with open('mappings.csv', 'wb') as outf:
        fieldnames = ['uri'] + ['category']
        csvwriter = csv.DictWriter(outf, fieldnames)

        while productCount > 0:
                row = {}
                csvwriter.writerow(dict(row, uri="/products/product%d" % productCount,
                category=random.choice(productCategories)))
                productCount -= 1
```

As you can see, there are 32 categories. You will execute the scripts as shown in Listing 6-2.

Listing 6-2. Mapping file extract

```
python mapping_generator.py
```

You should get following mappings.csv file:

```
...
/products/product90956,Shoes
/products/product90955,Shoes
/products/product90954,baby
/products/product90953,household applicance
/products/product90952,Books
/products/product90951,Audio and musique
/products/product90950,IT
/products/product90949,Devices accessories
/products/product90948,Devices accessories
/products/product90947,Jewellery
/products/product90946,car
/products/product90945,Devices accessories
/products/product90944,Toys and games
/products/product90943,baby
/products/product90942,Books
/products/product90941,IT accessories
/products/product90940,cycling
...
```

On the other side, the https://raw.githubusercontent.com/bahaaldine/elasticsearch-clickstream-demo/master/generator/resources.txt file contains all of the possible product URIs.

In the end, Spark will be in charge to correlate the URI to the category based on the mapping file. So let's analyze the needed Spark code in the following class:

https://github.com/bahaaldine/spark-enrich-and-ml/blob/master/src/main/scala/org/apache/spark/examples/SparkEnricher.scala

We kept the Clickstream object but added a Customer object that will be the output of our streaming.

Listing 6-3. Customer object

```
case class Customer (
    session:String,
    request:String,
    category:String
)
```

Note that the entire Spark project can be found on the following repo:

https://github.com/bahaaldine/scalable-big-data-architecture/tree/master/chapter6/spark-enrich-and-ml

The customer object is pretty simple so far and contains only three fields:

- A session field, which is the key to identify the user
- A request field, which contains the product URI
- A category field, which will be populate with the product category using the mapping file.

The Spark context configuration is unchanged; we just added another data source, our mapping file, as can be seen in Listing 6-4.

Listing 6-4. Loading the mapping file

```
val mappingFile = ssc.sparkContext.textFile("/path/to/mappings.csv")
val mappings = mappingFile.map(line => line.split(",")).map(x => (x(0),x(1))).collectAsMap()
val broadcastMapping = ssc.sparkContext.broadcast(mappings)
```

Remember that the previous Python script generates a mapping file called mappings.csv. You must set this path in this code to make the Spark drive works.

The mapping is loaded into a mappings variable, which is then broadcast to create a read-only variable put in the cache of each Spark machine. The variable is then used to create and enrich the Customer object, as seen in Listing 6-5.

Listing 6-5. Creating & Enriching the customer object

```
val customers = events.map { clickstream =>
  val lookupMap = broadcastMapping.value
  Customer(
    clickstream.clientip
    , clickstream.request
    , lookupMap.getOrElse(clickstream.request, "category not found")
  )
}
```

The mapping is loaded in the lookupMap variable, and we simply look up the product category by using the getOrElse. This returns "category not found" if the mapping does not exist. Finally, we index each customer object in Elasticsearch with the snippet shown in Listing 6-6.

Listing 6-6. Customer RDD indexation

```
customers.foreachRDD{ rdd =>
  if (rdd.toLocalIterator.nonEmpty) {
    EsSpark.saveToEs(rdd, "spark/customer")
  }
}
```

Each customer is indexed in the Spark index under the customer type in Elasticsearch. To test our code, make sure that the lambda architecture is fully running and launch the build.sh file that builds and launches our driver.

Listing 6-7. Spark driver launch command

```
./build.sh
```

By generating a new clickstream log with the generator.py script, you should get Customers indexed in the spark index. Just browse `http://elasticsearch_host:9200/spark/_search` to get the output shown in Listing 6-8.

Listing 6-8. Indexed customer in Elasticsearch

```
hits: {
total: 10,
max_score: 1,
hits: [
        {
                _index: "spark",
                _type: "customer",
                _id: "AU9-8BCuIm6jDjOublOw",
                _score: 1,
                _source: {
                        session: "86.192.183.220",
                        request: "/products/product106871",
                        category: "Watches"
                }
        },
        {
        _index: "spark",
        _type: "customer",
        _id: "AU9-8BCuIm6jDjOublO1",
        _score: 1,
                _source: {
                        session: "86.192.149.103",
                        request: "/products/product158310",
                        category: "cycling"
                }
        }
...
```

At this point we have an index full of mapped customers. We'll now leverage this enrichment phase to cluster our data and build a prediction.

Labelizing the Data

To train our model, we'll build a two-dimensional vector, which is created based on a category label and a customer label. The category label is associated to a specific category such as Car, Sport, and so on, whereas with a customer label, we'll introduce the fake concept of a customer type.

The customer label goes from one to five and can be understood as follows:

1. free

2. early bird

3. gold

4. platinium

5. big boy

117

So before training the model we need to create a couple of mapping files to be able to identify a category by a label and a customer by a label.

The https://github.com/bahaaldine/elasticsearch-clickstream-demo/blob/master/generator/mapping_generator.py has been slightly modified to add a label column, as seen in Listing 6-9.

Listing 6-9. Adding label to categories

```
category = random.choice(productCategories)
csvwriter.writerow(dict(row, uri="/products/product%d" % productCount, category=category,
label=productCategories.index(category)+1))
```

The label is calculated based on the position of the related category in the productCategories list variable. After executing the Python script, you should obtain a mapping file like that shown in Listing 6-10.

Listing 6-10. Category mapping file with label

```
...
/products/product199988,High-Tech accessories,2
/products/product199987,House,22
/products/product199986,baby,6
/products/product199985,Heatlh,31
/products/product199984,Watches,30
...
```

For the customer, I've created a new Python script located here:

https://github.com/bahaaldine/elasticsearch-clickstream-demo/blob/master/generator/ip_mapping_generator.py,

which simply adds a random integer between 1 to 5 next to each IP in the generated file, as seen in Listing 6-11.

Listing 6-11. IP mapping file generator

```
#!/usr/bin/python
import csv
import random

with open('ip_mappings.csv', 'wb') as outf:
        fieldnames = ['ip'] + ['label']
        csvwriter = csv.DictWriter(outf, fieldnames)

        for i in range(0,255):
                for j in range(0,255):
                        row = {}
                        csvwriter.writerow(dict(row, ip="86.192."+str(i)+"."+str(j),
                        label=random.randint(1,5)))
```

This sounds like a weird choice to use the IP as the identifier of the customer, but I did this on purpose for easing the comprehension of this part of the book and working on a finite range of customer identifiers.

After executing the Python script, you should obtain an equivalent output to that shown in Listing 6-12.

Listing 6-12. Ip mapping with label

```
...
86.192.0.15,3
86.192.0.16,3
86.192.0.17,5
86.192.0.18,3
86.192.0.19,1
86.192.0.20,4
86.192.0.21,1
...
```

This means that customer 86.192.0.21 is a free customer.

Training and Making Prediction

In this last section, we are going to use all of the asset already discussed to build a K-mean streaming machine learning driver. To do so, we'll reuse the previous Scala code and enrich it with the following steps.

First, we load the mappings files we just modified or created, as seen in Listing 6-13.

Listing 6-13. Loading the mapping file

```scala
val productCategoryMappingFile = ssc.sparkContext.textFile("/path/to/mappings.csv")

val productCategoryMapping = productCategoryMappingFile.map(line => line.split(",")).map(x
=> (x(0),x(1))).collectAsMap()

val categoryLabelMapping:scala.collection.Map[String,Double] = productCategoryMappingFile.
map(line => line.split(",")).map(x => (x(1),x(2).toDouble)).collectAsMap()

val brodcastProductCategoryMapping = ssc.sparkContext.broadcast(productCategoryMapping)

val brodcastCategoryLabelMapping = ssc.sparkContext.broadcast(categoryLabelMapping)

val customerMappingFile = ssc.sparkContext.textFile("/path/to/ip_mappings.csv")

val ipLabelMapping:scala.collection.Map[String,Double] = customerMappingFile.map(line =>
line.split(",")).map(x => (x(0),x(1).toDouble)).collectAsMap()
val brodcastIpLabelMapping = ssc.sparkContext.broadcast(ipLabelMapping)
```

Here we create three broadcasted maps:

- the product category map that we already know

- the category label map that is extracted from the modified mappings.csv

- the IP label mapping file that was extracted from the newly created ip_mappings.csv

The enrichment process remains unchanged; however, I've added two closures to create the training and the test data required for our model to make predictions, as shown in Listing 6-14.

Listing 6-14. Training data creation

```
val trainingData = customers.map { customer =>
  val categoryLookupMap = brodcastCategoryLabelMapping.value
  val customerLookupMap = brodcastIpLabelMapping.value

  val categoryLabel = categoryLookupMap.getOrElse(customer.category, 1).asInstanceOf[Double]
  val customerLabel = customerLookupMap.getOrElse(customer.session, 1).asInstanceOf[Double]

  Vectors.dense(Array(categoryLabel, customerLabel))
}
```

This first closure uses the category label map and the IP label map and build a two-dimentional vector that make the connection between a category and a customer type. For example, a vector can have the following value:

[30.0, 3.0]

This vector links the 29th (29+1, see mapping generator) category, which is "Shoes", to the third customer type, meaning gold customers.

The test data side is shown in Listing 6-15.

Listing 6-15. Testing data creation

```
val testData = customers.map { customer =>
val categoryLookupMap = brodcastCategoryLabelMapping.value
val customerLookupMap = brodcastIpLabelMapping.value

val categoryLabel = categoryLookupMap.getOrElse(customer.category, 1).asInstanceOf[Double]
val customerLabel = customerLookupMap.getOrElse(customer.session, 1).asInstanceOf[Double]

LabeledPoint(categoryLabel, Vectors.dense(Array(categoryLabel, customerLabel)))
}
```

Here we create a different kind of object, a LabelPoint. This object has two members:

- some label: here I've set the same label than the category
- a vector: the vector is build in the same way than in the training data

Finally, we creat and train our model, as seen in Listing 6-16.

Listing 6-16. Create and train the model

```
val model = new StreamingKMeans()
  .setK(3)
  .setDecayFactor(1.0)
  .setRandomCenters(2, 0.0)

model.trainOn(trainingData)
```

We specify here that we need three centroids, set a decay factor because we are streaming data, and note that they are not linear. We need the model to train on the most recent data with a reflection of the past data.

Then we call the model to train against the training data. We just have now to call the predicted method and obtain the expected result, as seen in Listing 6-17.

120

Listing 6-17. Trigger prediction

```
model.predictOnValues(testData.map(lp => (lp.label, lp.features))).foreachRDD{ rdd =>
  if (rdd.toLocalIterator.nonEmpty) {
    EsSpark.saveToEs(rdd, "spark/prediction")
  }
}
```

You should now see prediction data coming in Elasticsearch, ready to be leveraged for further recommendation engine or simply analytics, as seen in Listing 6-18.

Listing 6-18. Prediction indexed in Elasticsearch

```
...
{
        _index: "spark",
        _type: "prediction",
        _id: "AU-A58AEIm6jDjOublw2",
        _score: 1,
        _source: {
                _1: 24,
                _2: 2
        }
},
{
        _index: "spark",
        _type: "prediction",
        _id: "AU-A58AEIm6jDjOublw7",
        _score: 1,
        _source: {
                _1: 8,
                _2: 1
        }
}
...
```

All you need now is a little bit of imagination to think of what you can do with this data to improve the user experience of your website visitors. You can build a recommendation engine on top of that, to push a banner, in-app notifications, and emails to "early-bird" customers about category 23th (24+1, see mapping generator), which is "Music".

Summary

Our architecture is getting more and more complex. We have seen in this chapter how we can enrich our data processing pipeline with a machine learning feature and cluster our data as they come in Elasticsearch.

The next chapter will deal with governance and main assets that should be set up to reduce risks in production.

■ ■ ■

## CHAPTER 7

Governance Considerations

One main aspect of the architecture is its governance. In other words, how will it be sized, deployed, or event monitored? In this chapter, I'll go through governance considerations starting with the "Dockerization" of the architecture, while talking about scalability and its impact on the architecture.

Dockerizing the Architecture

There are different ways to deploy a data architecture; you can install components on distributed physical machines or on virtual machines. In the past few years, Docker has become a standard for distributed architecture deployment, so you'll use it in this book for the example use case.

Introducing Docker

Docker is an open source platform that allows users to reduce the deployment cycle time between the development stage and the production stage. It provides a comprehensive set of tools and APIs that will help you to build an infrastructure that will run your application.

For the use case in this book, you'll be able to deploy the full architecture using a Docker container. Docker has its own terminology, defined in the following list:

- *Docker container*: This is the smallest working unit that runs your application. For example, you can have a Docker container that runs an Apache Tomcat server to serve your RESTful API.

- *Docker image*: This is an image that describes a Docker container. So, in this example, you will need an operating system and some packages to run Tomcat. All the needed assets and configuration are written in a read-only template called a Dockerfile, which is the Docker image. For more information on the Docker image, see `https://docs.docker.com/userguide/dockerimages/`.

- *Docker client*: This is the first user interface to interact with the underlying Docker features. Once your Docker image is written, you will build it to run a Docker container. This is something you will do using the Docker client. For more information on the Docker command line, see `https://docs.docker.com/reference/commandline/cli/`.

- *Docker daemon*: This orchestrates all the underlying operations needed to run the Docker containers. The user interacts with the daemon through the Docker client.

Figure 7-1 summarizes what I have just described.

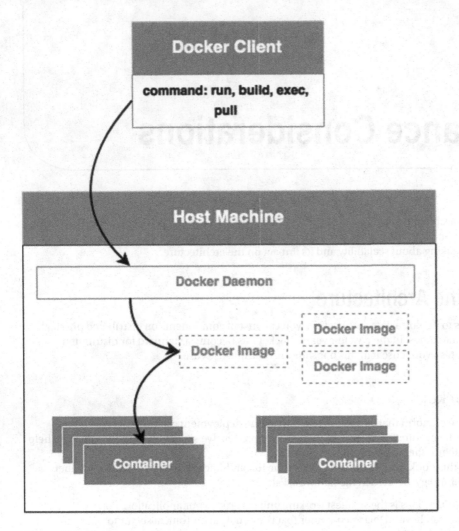

Figure 7-1. *Docker components*

So, in the example use case, if you focus on the data-processing part that involves multiple technologies, Docker is a great benefit for running and provisioning the infrastructure. You basically just have to create a Docker image for every technology that needs to run on a different container, meaning for the following:

- Logstash forwarder

- Logstash, processor, and indexer are using the same technology, but only a different configuration

- Kafka

- Zookeeper

- Elasticsearch

- Kibana

Installing Docker

Docker supports a multitude of configurations for installation. In the scope of this book, I'm running Docker on Mac OS X that has a slightly different installation steps than on Ubuntu, for example.

I recommend you visit the following page and go through the relevant installation process for your system:

`https://docs.docker.com/installation/`

You will also need Docker Compose, which you can get here:

`https://docs.docker.com/compose/install/`

You will need Docker machine, which you can get here:

`https://docs.docker.com/machine/install-machine/`

At the end of the process, make sure you can run Docker and the Docker Compose command line by issuing the code in Listing 7-1.

Listing 7-1. Docker Installation Checks

```
bash-3.2$ docker -v
Docker version 1.7.1, build 786b29d

bash-3.2$ docker-compose -v
docker-compose version: 1.3.2
CPython version: 2.7.9
OpenSSL version: OpenSSL 1.0.1j 15 Oct 2014

bash-3.2$ docker-machine -v
docker-machine version 0.3.0 (0a251fe)
```

In this code, you can see the version of Docker components I'm working with while writing this book. At this point, you are ready to start creating your infrastructure Docker images.

Creating Your Docker Images

In this section, I'll explain in detail how to create one of your architecture application's Docker image. You will see that the overall procedure is the same for every other application, so I'll describe the creation of just one.

For reference, you can find all the images on my GitHub repositories.

- *Elasticsearch*: `https://github.com/bahaaldine/docker-elasticsearch`
- *Logstash*: `https://github.com/bahaaldine/docker-logstash`
- *Logstash Forwarder*: `https://github.com/bahaaldine/docker-logstash-forwarder`
- *Kibana*: `https://github.com/bahaaldine/docker-kibana`
- *Zookeeper*: `https://github.com/bahaaldine/docker-zookeeper`

- *Apache Kafka*: https://github.com/bahaaldine/docker-kafka
- *Apache Zookeeper for Kafka*: https://github.com/bahaaldine/docker-zookeeper
- *Apache Spark*: https://github.com/bahaaldine/docker-spark

Let's now focus on creating an Elasticsearch Docker image and walking through the Dockerfile file, available here:

https://github.com/bahaaldine/docker-elasticsearch/blob/master/Dockerfile

The template is divided into several parts, each responsible for a specific job.

- *The image configuration*: You start by setting the container type, which is a Java container here, that will run Elasticsearch, and you set a couple of description metadata tags. Here is the maintainer:

```
FROM java:8
MAINTAINER Bahaaldine Azarmi baha@elastic.co
```

- *The OS-level configuration*: Then you create the OS-level configuration by setting the needed environment variable, upgrading your container package list, and installing the needed packages.

```
ENV DEBIAN_FRONTEND noninteractive
RUN apt-get update
RUN apt-get install -y supervisor curl

# Elasticsearch
RUN \
    apt-key adv --keyserver pool.sks-keyservers.net --recv-keys
    46095ACC8548582C1A2699A9D27D666CD88E42B4 && \
    if ! grep "elasticsearch" /etc/apt/sources.list; then echo "deb
    http://packages.elasticsearch.org/elasticsearch/1.6/debian stable main" >>
    /etc/apt/sources.list;fi && \
    apt-get update
```

- *The application-level installation*: Installation is pretty and just requires you to run the following command:

```
apt-get install -y elasticsearch
```

- *The application configuration*: You then run a couple of sed commands to modify the Elasticsearch configuration file (elasticsearch.yml) to set your cluster configuration. In the following code, you are setting the cluster name, the path to the directory where Elasticsearch will store its data, the number of shards and replicas, and a network configuration for publishing the proper host:

```
sed -i '/#cluster.name:.*/a cluster.name: logging'
/etc/elasticsearch/elasticsearch.yml && \
sed -i '/#path.data: \/path\/to\/data/a path.data: /data'
/etc/elasticsearch/elasticsearch.yml && \
```

```
sed -i '/#index.number_of_shards:.*/a index.number_of_shards: 1'
/etc/elasticsearch/elasticsearch.yml && \
sed -i '/#index.number_of_replicas:.*/a index.number_of_replicas: 0'
/etc/elasticsearch/elasticsearch.yml && \
sed -i '/#marvel.index.number_of_replicas:.*/a index.number_of_replicas: 0'
/etc/elasticsearch/elasticsearch.yml && \
sed -i '/#network.publish_host:.*/a network.publish_host: elasticsearch'
/etc/elasticsearch/elasticsearch.yml
```

- *The application runtime configuration*: You use Supervisor (http://supervisord. org/), a process control system, here to orchestrate the execution of the applications installed in the container. So, you need to create the configuration to make sure it works. Then you install Marvel, an Elasticsearch monitoring plug-in, and expose the 9200 port to the container outside.

```
http://supervisord.org/ADD etc/supervisor/conf.d/elasticsearch.conf /etc/
supervisor/conf.d/elasticsearch.conf
```

```
RUN /usr/share/elasticsearch/bin/plugin -i elasticsearch/marvel/latest
```

```
EXPOSE 9200
```

- *The container runtime configuration*: The last step is to ask Supervisor to run using the Supervisor configuration file.

```
CMD [ "/usr/bin/supervisord", "-n", "-c", "/etc/supervisor/supervisord.conf" ]
```

If you check the other Docker images, you will see that a few are changing from one image to the another, as I'm always using these steps to create my Docker images.

You have to build a Docker image for it to be used. To build the Elasticsearch image, run the following command in the Docker image folder:

```
docker build -t bahaaldine/docker-elasticsearch .
```

The build command will search for a Dockerfile file in the current directory and start building a Docker image called bahaaldine/docker-elasticsearch (although you can change the name if you'd like); in any case, you can check your local image repository by issuing the following image and see whether your image is there:

```
bash-3.2$ docker images
REPOSITORY    TAG    IMAGE ID    CREATED    VIRTUAL SIZE
bahaaldine/docker-elasticsearch    latest    a3422c874201    About a minute ago    877.4 MB
```

Now to spin up your container, run the following command:

```
bash-3.2$ docker run -ti -p 9200:9200 -p 9300:9300 --add-host="elasticsearch:192.168.59.103"
bahaaldine/docker-elasticsearch
```

```
2015-08-23 13:33:24,689 CRIT Supervisor running as root (no user in config file)
2015-08-23 13:33:24,689 WARN Included extra file "/etc/supervisor/conf.d/elasticsearch.conf"
during parsing
2015-08-23 13:33:24,697 INFO RPC interface 'supervisor' initialized
2015-08-23 13:33:24,697 CRIT Server 'unix_http_server' running without any HTTP
authentication checking
2015-08-23 13:33:24,697 INFO supervisord started with pid 1
2015-08-23 13:33:25,701 INFO spawned: 'elasticsearch1' with pid 7
2015-08-23 13:33:27,331 INFO success: elasticsearch1 entered RUNNING state, process has
stayed up for > than 1 seconds (startsecs)
```

The Elasticsearch container is now running and exposes ports 9200 and 9300 on same ports on the host machine, which means that if you try to connect to 192.168.59.103:9200, you should get the Elasticsearch REST API response, as shown in Listing 7-2.

Listing 7-2. Elasticsearch Installation Check

```
{
    status: 200,
    name: "lambda-architecture",
    cluster_name: "logging",
    version: {
        number: "1.6.2",
        build_hash: "622039121e53e5f520b5ff8720fdbd3d0cb5326b",
        build_timestamp: "2015-07-29T09:24:47Z",
        build_snapshot: false,
        lucene_version: "4.10.4"
    },
    tagline: "You Know, for Search"
}
```

A couple of questions came to my mind the first time I was trying to Dockerize my architecture: What if I have a ten-node cluster? What about the rest of the architecture? Will I have to run every container with a specific configuration at the command line?

You clearly understand the benefit of Dockerizing your architecture application, but it would be a nightmare to manage it if you have to take of each container. That's where Docker Compose comes into play.

Composing the Architecture

Composing the architecture has a double meaning here; it first means that you would like to combine Docker containers to build the architecture, but it also means you need a high-level tool that will help you to govern each container configuration and runtime preferences.

Fortunately, Docker Compose is the perfect tool for this duty. Compose is based on a single-file configuration that will allow users to create multicontainer applications in a declarative way. While Compose is not recommended for production, it's essential in development, staging, or continuous integration (CI).

In this section, I'll walk you through the Docker Compose file located at the following GitHub project:

`https://github.com/bahaaldine/scalable-big-data-architecture/tree/master/chapter7/docker`

Like with the images, you'll focus on one container configuration, and you'll see that the configuration is pretty much the same for the others. The compose file is spinning up an architecture that counts the following:

- A two-node Kafka cluster with a one-node Zookeeper cluster
- A one-node Elasticsearch cluster
- A single instance of Kibana
- One Logstash forwarder
- One Logstash processor
- One Logstash indexer

Listing 7-3 shows the result of the composition.

Listing 7-3. Docker Compose Lambda Architecture File

```
zookeeper1:
  image: bahaaldine/docker-zookeeper
  volumes:
    - "conf/kafka/server1:/etc/kafka"
    - "logs/zookeeper1:/opt/kafka/logs/"
  ports:
    - "2181:2181"
  extra_hosts:
    - "brokers:192.168.59.103"

kafka1:
  image: bahaaldine/docker-kafka
  volumes:
    - "conf/kafka/server1:/etc/kafka"
    - "logs/kafka1:/opt/kafka/logs/"
  extra_hosts:
    - "brokers:192.168.59.103"
  ports:
    - "9092:9092"

kafka2:
  image: bahaaldine/docker-kafka
  volumes:
    - "conf/kafka/server2:/etc/kafka"
    - "logs/kafka2:/opt/kafka/logs/"
  extra_hosts:
    - "brokers:192.168.59.103"
  links:
    - "kafka1"
  ports:
    - "9093:9093"
```

```
logstashProcessor1:
  image: bahaaldine/docker-logstash-agent
  volumes:
    - "conf/logstash/processor:/etc/logstash"
    - "conf/security:/etc/logstash/security"
    - "logs/logstash-processor1:/var/log/logstash"
  links:
    - kafka1
  ports:
    - "5043:5043"

elasticsearch1:
  image: bahaaldine/docker-elasticsearch
  ports:
    - "9200:9200"
  volumes:
    - "logs/elasticsearch1:/var/log/elasticsearch"
    - "conf/elasticsearch/templates:/etc/elasticsearch/templates"
    - "data:/data"
  extra_hosts:
    - "elasticsearch:192.168.59.103"

logstashIndexer1:
  image: bahaaldine/docker-logstash-agent
  volumes:
    - "conf/logstash/indexer:/etc/logstash"
    - "logs/logstash-indexer1:/var/log/logstash"
  links:
    - elasticsearch1
  extra_hosts:
    - "brokers:192.168.59.103"
    - "elasticsearch:192.168.59.103"

logstashForwarder:
  image: bahaaldine/docker-logstash-forwarder
  volumes:
    - "conf/logstash/forwarder:/etc/logstash-forwarder"
    - "conf/security:/etc/logstash-forwarder/security"
    - "logs/logstash-forwarder1:/tmp/logs/"
    - "source:/tmp/source"
  extra_hosts:
    - "processors:192.168.59.103"
kibana1:
  image: bahaaldine/docker-kibana
  ports:
    - "5601:5601"
  volumes:
    - "logs/kibana:/var/log/kibana"
  extra_hosts:
    - "elasticsearch:192.168.59.103"
```

If you now focus on the Elasticsearch container configuration, you might recognize some of the parameters you passed at the command line earlier when you tried to run your container.

```
docker run -ti -p 9200:9200 -p 9300:9300 --add-host="elasticsearch:192.168.59.103"
bahaaldine/docker-elasticsearch
```

What you can do with the Docker run command can be described in the Compose YML file.

1. You first set the container image.

   ```
   image: bahaaldine/docker-elasticsearch
   ```

2. Then you map the host port to the container port.

   ```
   ports:
     - "9200:9200"
   ```

3. Next you map the host file system with container directories for the logs, configuration, and data stored in Elasticsearch. This is important because when a container is shut down, all data stored inside the container is lost. So, you are mapping those folders to make sure they will live beyond the container life cycle.

   ```
   volumes:
     - "logs/elasticsearch1:/var/log/elasticsearch"
     - "conf/elasticsearch/templates:/etc/elasticsearch/templates"
     - "data:/data"
   ```

4. Last, you ask Docker to add a host to the /etc/hosts file.

   ```
   extra_hosts:
     - "elasticsearch:192.168.59.103"
   ```

That's it. As I mentioned earlier, every other container is configured in the same way; you just have to run the command shown in Listing 7-4 to start the architecture.

Listing 7-4. Docker Compose Project Start Command

```
docker-compose up
```

Note that here I've built all the images before launching the Compose project. This is because I've pushed all the images on the public Docker registry, which is called the Docker Hub (https://hub.docker.com/). If the images were not available, you would get an error.

The My Compose project comes with the generator script, which should be used to create some data in the source directory. Remember that the Logstash-forwarder is transporting files from this directory; this also appears in the Compose YMP file, as shown in Listing 7-5.

Listing 7-5. CFC

```
logstashForwarder:
  image: bahaaldine/docker-logstash-forwarder
  volumes:
    - "conf/logstash/forwarder:/etc/logstash-forwarder"
    - "conf/security:/etc/logstash-forwarder/security"
    - "logs/logstash-forwarder1:/tmp/logs/"
    - "source:/tmp/source"
```

The source directory is mapped to the container's /tmp/source folder.

From a governance point of view, Docker eases the process of provisioning your architecture; you now need to focus on the architecture scalability and how you can size it in order to scale out if needed.

Architecture Scalability

In this section, I'll talk about the different scalability zones that exist in the data architecture and make some recommendations.

Sizing and Scaling the Architecture

When you are sizing your architecture to handle a certain data volume and reach certain SLAs, xx. The answer is always it depends.

The goal in this section is not to give a magic recipe or formula that provides every part of your architecture with the exact sizing it needs in terms of hardware but more to give you directions so you can better handle this question.

The sizing of the architecture is needed at different levels of your architecture:

- At the logstash processing and indexing level

- At the Kakfa/Zookeeper level

- At the Elasticsearch level

For all these layers, there are common aspects to consider to create the right sizing.

- *The size of input data per day, and so on*: Knowing how data will move through the architecture will help you estimate the amount of store that Elasticsearch will need. Based on the total size per node, you can estimate the size of the indexed data without forgetting to apply a compression rate; for example, a basic raw log would end up as an indexed JSON document that is bigger in terms of size.

- *The structure of the ingested data*: The data can have a simple structure such as Apache logs or multiple-line structure such as an application exception in Java.

- *The average number of events ingested per second*: This gives information on the memory and the CPU needed for each layer in your architecture to support the load.

- *The retention period*: When ingesting time-series data, sometimes it's important to keep in the cluster data that is relevant for search, even if it starts to be old. That's the principle of retention, and this might cost a lot of disk space depending on the needed period.

- *If high availability is needed*: In most enterprise use cases, you want to make sure that if an instance of an application fails, the overall architecture keeps running. Each application might have different terminology, but that's the principle of resiliency. Add more servers, replicas, and instances to avoid data loss; service continuity brings hardware cost because obviously you don't want to have several instances of your application on the same machine.

You also have some specific sizing considerations for each application.

- *The expected indexing throughput*: Data ingestion and indexation can have an uneven pace, meaning you might have different load depending on the period of the day, such as on e-commerce web site. It's important then to set the SLA that secures those kinds of uncommon but significant activities.

- *The number of dashboards in Kibana*: Remember that Kibana relies on the Elasticsearch API to build the visualizations that compose a dashboard. So, whenever you refresh a Kibana, it sends a request to Elasticsearch and generates a search or aggregation load that the cluster might consider if it's significant.

- *The number of visualization users*: Depending on the use case, you might have a different number of users who will connect to Kibana to view the dashboard. This will have an effect on the sizing.

Having all this information is not enough; you also need to experiment and analyze the KPI I just talked about to see whether your architecture is sustainable. A naïve but efficient way to do that is to start with a pseudodistributed mode architecture, in other words, a single instance of each application across the architecture. This rule doesn't apply to Zookeeper, which needs at least three servers to be resilient; the same is true for Elasticsearch master nodes that are specific nodes dedicated to cluster state management.

This experiment can be applied to Kafka, Elasticsearch data nodes, and Logstash processors/indexers. Once you get the metrics for each layer on a single-instance mode, scale out linearly and apply a performance test again.

Other parameters can influence the sizing such as the number of replicas on Elasticsearch that augment resiliency but also search speed in a lot of use cases. Adding replicas increases the number of needed machines and thus your sizing.

Figure 7-2 shows that the Lambda architecture has multiple scalability zones and scales out, meaning that adding more nodes to an application to add resilient and load support is pretty easy.

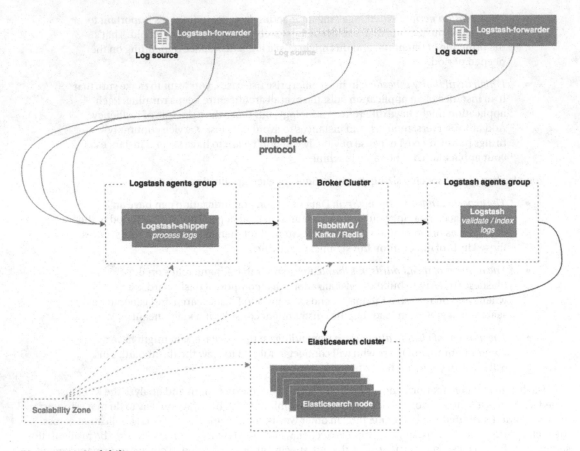

Figure 7-2. *Scalability zone*

The point of this section is to go through the considerations when scaling out a distributed architecture.

From a provisioning point of view, since the architecture is Dockerized, it is easy to scale because you just need to add new containers with related configuration.

At the ingestion level, there are no notions of the Logstash cluster so far; each instance consists of lightweight runtimes that scale independently. So, as you scale out, the main thing to do is to configure the same input to parallelize the workload.

At the Elasticsearch level, when a document is indexed, it's stored in a single primary shard. Elasticsearch evaluates in what shard it should index the document based on the formula in Listing 7-6.

Listing 7-6. Document Routing Formula

```
shard = hash(routing) % number_of_primary_shards
```

So, adding nodes in the cluster won't affect anything and may give a better shard distribution; however, adding more primary shards will invalid the previous routing value and will need to re-index all data to be retrieved. This starts to be an issue when you deal with a large amount of data that takes a lot of time to re-index.

Monitoring the Infrastructure Using the Elastic Stack

In previous chapters, you learned to set up a Lambda architecture that ingests an Apache log to reveal business analytics. Each layer of the architecture is responsible for a specific task in the overall transaction, and there may be different important KPIs per application to consider monitoring.

The question is, why don't you scale the acquired knowledge to monitor your own architecture? The idea behind this question is to enable your architecture log to be ingested for further operational intelligence.

The only thing you need to do is organize your architecture with the following steps:

1. Centralize your logs. If the logs are centralized in one place such as a SAN, an Amazon S3 bucket, or another network share folder, that will ease the retrieval of logs information and also the deployment of the monitoring architecture. In the Docker Compose file, the configuration is set to centralize every log in the Docker Compose project logs folder. Listing 7-7 shows the configuration for Zookeeper.

Listing 7-7. Docker Compose Kafka Configuration

```
kafka1:
  image: bahaaldine/docker-kafka
  volumes:
    - "conf/kafka/server1:/etc/kafka"
    - "logs/kafka1:/opt/kafka/logs/"
  extra_hosts:
    - "brokers:192.168.59.103"
  ports:
    - "9092:9092"
```

2. Then you should consider reusing your existing Elasticsearch cluster or creating another one. Obviously, the answer to this consideration will depend on different factors, such as the density and workload of the existing cluster, or even nontechnical factors such as financial considerations that can occur when it comes time to acquire a new machine. If it's a dedicated cluster, then it's easier; if not, the events should simply be stored in a new index, for example behind a new "monitoring" index alias.

   Sometimes different event types are indexed; you should remember that Kibana will deal with the full index without making distinctions of different types of documents within the index. So, naming your event properly so there are no field collisions is also a consideration that you should keep in mind.

3. Using Curator (https://github.com/elastic/curator) to manage your indices is essential too! I talked about the retention period earlier, which is also relevant for monitoring your architecture. As monitoring logs are ingested and indexed, time-based indexes are created, and with Curator, you will be able to schedule procedures that will close, archive, or delete indexes (for example, every index older than six weeks).

The last point to consider is the security of the overall architecture, as described in the next section.

Considering Security

Security is complex, specifically in distributed architecture. Ideally there are a set of tools and best practices that reduce the pain of securing the architecture.

- The first tip is related to Dockerizing your architecture. Every container is isolated from the rest of the world. You can limit the number of resources that the container will use such as the RAM, CPU, or network usage, but you can also limit access in terms of network connectivity.

- Still, sometimes you need to expose the container to the outside world. For instance, in the previous example, I set the configuration of the Docker Compose YML file so that the Elasticsearch container is exposing port 9200, as shown in Listing 7-8.

Listing 7-8. 8CC

```
elasticsearch1:
  image: bahaaldine/docker-elasticsearch
  ports:
    - "9200:9200"
  volumes:
    - "logs/elasticsearch1:/var/log/elasticsearch"
    - "conf/elasticsearch/templates:/etc/elasticsearch/templates"
    - "data:/data"
  extra_hosts:
    - "elasticsearch:192.168.59.103"
```

- This represents a security breach if the port is not secured on the host machine. Ideally, this doesn't require any specific security configuration and can be handled by the usual security best practices such as IP filtering, proxies, and so on.

- The log transport from the forwarder to the processor is secured through the Lumberjack protocol that uses certificates.

- Kibana also needs to expose its port to give users access to the dashboards. The same applies here in terms of security best practices; you can simply deploy a proxy web server such as an Apache web server and configure it to expose an HTTPS endpoint.

- Last is accessing the data in Elasticsearch. Even if you put a proxy in front of Elasticsearch, one user can access the other user's indices by default. As of today, the only supported way to secure the data and access to Elasticsearch is by using one of the commercial plug-ins called Shield.

Shield is an Elasticsearch plug-in that offers multiple security features for the ELK stack, including the following:

- *Role-based access controls*: Roles can be defined and mapped to users or groups of users. Every privilege can be finely configured such as cluster and indices-level privileges. In this case, you can segregate a shared Elasticsearch cluster so every user will see only the data and assets to which he has related privileges.

- *LDAP or Active Directory users*: Shield can connect to your LDAP or Active Directory so you don't have to redefine all your users at Elasticsearch; this affects only roles, and you can leverage your existing authentication mechanism. Users can be actual human users or even technical users such as a Logstash agent that indexes documents in Elasticsearch.

- *Encrypted communications*: Shield allows encrypted communications through SSL and TLS.

- *Audit logging*: Every access to the cluster is logged.

- *IP filtering*: You can restrict access to specific IP address or a range of IP addresses.

So, as the cluster grows, offers more entry points to users, and shares resources between different data projects, you might consider adding an enterprise plug-in such as Shield to secure your architecture.

Summary

The data architecture is by nature distributed and highly scalable. In this chapter, you learned how you can keep bringing flexibility to the governance aspect by packaging the architecture with Docker Compose. At this point, you should be able to extend this architecture by, for example, adding new components and enriching the Docker Compose project with new configurations.

Index

■ U, V, W, X, Y, Z

Get the eBook for only $5!

Why limit yourself?

Now you can take the weightless companion with you wherever you go and access your content on your PC, phone, tablet, or reader.

Since you've purchased this print book, we're happy to offer you the eBook in all 3 formats for just $5.

Convenient and fully searchable, the PDF version enables you to easily find and copy code—or perform examples by quickly toggling between instructions and applications. The MOBI format is ideal for your Kindle, while the ePUB can be utilized on a variety of mobile devices.

To learn more, go to www.apress.com/companion or contact support@apress.com.

Printed in the United States
By Bookmasters